A Time Warner Company

WARNER BOOKS EDITION

Cover design by Don Puckey
Cover photo by Image Bank

Warner Books, Inc.
666 Fifth Avenue
New York, N.Y. 10103

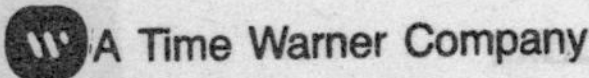
A Time Warner Company

Printed in the United States of America

First Printing: March, 1991

10 9 8 7 6 5 4 3 2 1

$2.50

BLOOD GURGLED FROM BETWEEN HER LIPS

•

The woman was still feebly moving in the spray of glass and drugs and blood strewn across the floor of the lab. Her hands were clawing at her face.

"The Others," Scarborough said between her moans, kneeling by her. "Who are the Others?"

She opened her eyes and there was nothing but fear and horror in them.

"Get away! Get away from me!" she gasped.

"The Others, woman. Who are they? Where are they? Where is my daughter! Who are the Others?"

"You are, Scarborough," she said in less than a voice, more of a croak. "You are..."

Also by
David Bischoff

ABDUCTION: THE UFO CONSPIRACY
RESOLUTION: THE UFO CONSPIRACY*

Published by
WARNER BOOKS

**forthcoming*

For
Steve Brown and Joanne Klappauf

ACKNOWLEDGEMENTS

Special thanks to Andy Zack, Cathy Solomon and of course, my editor, Brian Thomsen

A NOTE TO THE READER

Although this volume of THE UFO CONSPIRACY can be enjoyed on its own, it should be noted that it has a predecessor, ABDUCTION: THE UFO CONSPIRACY and will have a final and concluding volume, RESOLUTION: THE UFO CONSPIRACY.

PROLOGUE

Roswell, New Mexico July 3, 1947

Lieutenant Todd Jenkins of the United States Air Force was sitting in the mess hall drinking a cup of what the service claimed was coffee, doing the crossword puzzle from an out-of-date edition of *Stars and Stripes*, when a private came barrelling into the Quonset hut like a puffing bat out of hell.

"Lieutenant Jenkins, sir," said the private. Jenkins recognized him as Marousek, one of the communication geeks from the radio shack. "This is just incredible, sir!"

"Hold your horses and hang onto your drawers there, pal," said Jenkins, holding up a calming hand. "Unless the Russians just invaded Alaska, I think we can take a few seconds to catch our breaths."

Marousek took in a few gulps of air, but it only seemed to make his chubby face redder. "Sir, it's on the radio. We've got a report from the Roswell sheriff, about one of the local ranchers." Marousek nervously ran a hand through his short brown, already disheveled hair, and let the rest out in a gust of words. "Sir, it's a *flying disk*!"

Jenkins raised an eyebrow as he put down his cup of bitter coffee. Now, this was interesting. He didn't believe a word of it, but it was damned interesting nonetheless, and *interesting* wasn't a commodity one got much of here at the 509th Bomb Group of the Eighth Air Force, Roswell Army Air Field. Couple years ago, he thought he'd had his fill of action with Double-U Double-U Number Two. He'd served in both the European and the Pacific Theaters, merrily bombing the bejesus out of Kraut and Nip alike. But as a career officer still in his twenties, a whole lot of mountains, tumbleweeds, and rattlesnakes sure got boring after a while, even if you spent a lot of time flying above them.

"You mean a flying saucer?" he asked calmly, a hint of a smile dimpling his fair face below the blonde locks that still placed him in good stead with the ladies.

"Flying saucer . . ." Marousek repeated. "Yeah, okay. Flying saucer. Some farmer by the name of Brazel or something. Saw a light in the sky last night. This morning he wakes up to herd some cows or sheep or something, and he finds wreckage all over his range. And right in the middle of a field—a saucer. A *crashed* flying saucer. That's it. I don't know anymore, 'cept we got orders from the Eighth Air Force along with the report to go check it out. And we got to bring a flatbed and a tow truck, 'cause we're supposed to pick it up. Major Marcel says he's real busy this afternoon. He sent me over to you to take care of it."

"You want some coffee, Private?"

"Sir, this is priority." He lowered his voice when he realized that there was activity behind the counter. "And it's supposed to be *top* secret."

"Sit down for a second, friend. Let me finish my coffee here. Let me tell you something. How long you been in the forces?"

"Just a year, sir." Private Marousek sat down, looking antsy and uncomfortable, as though wondering which orders to obey, the ones from Fort Worth or the one from his superior here?

"I heard about these flying saucers in last Sunday's papers," said Lieutenant Jenkins. "That guy Kenneth Arnold, private pilot up in Washington State. You hear about it?"

"Yes, sir. It was the talk of the barracks. We were wondering if it's the Russians. Walnard—he reads a lot of pulp fiction magazines like *Weird Tales* and *Amazing*—talks a lot about stuff like the Shaver Mystery, and has an autographed picture of this dwarf named Raymond Palmer. Anyway, he says that they're either beings from another planet . . . or maybe even people from a civilization from the center of the earth."

Jenkins snorted. "Yeah, sure. My brother used to read Edgar Rice Burroughs's shit. Let me clue you in, Marousek, before you bust a vein or anything. When I was a copilot dumping metal over Strasbourg and Hamburg, I seen these 'flying saucers.' Lots of pilots seen 'em. Bit startling at first . . . but you get used to it."

Marousek's big, nervous eyes suddenly got bigger. "You *did*?"

"Hell, yeah! We used to call them 'FOO-fighters.' Big globs of lights, used to buzz around the wings and fuselage like flys around cow pies! Those and gremlins would drive a superstitious twerp nuts if you let it get to you—but hell, you realize pretty quick that what you gotta worry about are Spandaus and Zeros, not fairies. Anyway, a good talk with an army meteorologist set me straight. Ever hear of ball lightning? Yeah, well, lots of wierd electrical stuff goes on in the sky."

"But this isn't in the sky, sir. And it's not electrical. It's on the ground! Corporal Tankerslee's getting the trucks and he sent me over to get you. Really, sir, we gotta go!"

"Oh, sure, Private. We're gonna go, but let me finish my coffee first, huh? And I don't want you to be having no goddamned heart attack, that's all. Fuckin' credulous dipshit. You live through a war, you learn to take it easy, don't jump in quick or you'll get your goddamned tail shot off. Am I understood, Private Marousek?"

Private Marousek stared down glumly at the pile of crumbs on the table. "Yes, sir."

Jenkins sipped at his coffee. You had to show these young guys who was in control. Couldn't have one of them telling you what to do, or you'd lose their respect. So, even though Jenkins wasn't particularly *enjoying* this condensed milk and mud concoction in his mug, he was damned sight gonna finish it, just to spite everyone.

Moments of awkward silence passed.

"Anyway, Private," Jenkins said finally, relenting. "You know we send up weather balloons all the time. That's probably what it is. Popped weather balloon. You don't think this farmer didn't read this Sunday's paper, just like I did? You mark my words, that's what got this flying saucer nonsense in his bonnet!'' The lieutenant put his mug down, stood up and put his jacket and cap on slowly. Then he grinned down at Marousek. ''So what are we waiting for, Private! We gotta go check out these fuckin' *Martians*!''

It was a bright, clear day in Roswell, New Mexico. The air was clean and pure, excellent stuff for Lieutenant Jenkins's sinuses, which had been acting up ever since those years of poor pressurization in B-25 bombers. He sat shotgun in a beat-up old truck, Private Marousek driving. Three more privates were sitting in the covered back. They were following another truck—a Mack flatbed. Behind them was the requisitioned tow truck, with three more soldiers making a party of ten from the army field. Just to make sure, Lieutenant Jenkins had called up the Chavez County sheriff's office. Sure enough, the sheriff wasn't there, according to a deputy. He was out in some farmer's field, looking at some kind of wreckage. And yes, they'd contacted the Air Force.

The road was dirt, but it wasn't dusty since it had just rained the other day, and it had packed down some. The sun was up toward the middle of the sky, inching toward noon. The landscape was flat and filled with scrub brush for miles, with Haystack Mountain poking up among a stretch of fellow mountains in the distance.

''Can you believe that Corporal Narden?'' said Jenkins. ''Radiation suits! Christ, a wasted half hour getting those suckers out. Where does he think we're going, Hiroshima?''

''Los Alamos *is* in this state, sir.''

''Yeah? And three hundred miles away!'' Jenkins snorted with disgust. ''Sign of the times, I guess. Radiation madness. You mark my words, this thing in this farmer's field is about as radioactive as my ass!''

Private Marousek had nothing more to say. He'd apparently learned his lesson—when you had a lieutenant on the other end of a conversation, you didn't cross his opinions.

''Well, all I gotta say, Private,'' said Jenkins, half-talking to himself. ''I just hope this doesn't snowball into a rumor, like back in '38 in New Jersey.''

''There was a crashed flying saucer in New Jersey, sir?''

"Hell no. I was in diesel school outside Asbury that autumn. That Orson Welles guy... You know, the one who made that film *Citizen Kane*? Well, he had this radio dramatization on the radio of H.G. Wells's *War of the Worlds*. He made it sound like news reports, and people tuning in during the middle thought it was the real thing! Well, half the state was up and running away. Thought Martians were invading. Jesus. People are really something, you know, Marousek?"

The private nodded. "Yes, sir."

A tumbleweed whipped into their way and was immediately run down. This guy Brazell's ranch was twenty-five miles out into Nowhere! Lots of sky, lots of land, lots of nothing. God's country. Sheesh, thought Jenkins. Gimme the Jersey coast. Gimme New York. God may own this shit, sure; but dollars to donuts he lives in a penthouse above Fifth Avenue in Manhattan!

The military vehicles growled and ground along for what seemed like forever, until the promised signpost BRAZEL reared up ahead amidst the browns and brush sprawling arid land. "No fences!" said Jenkins. "How does this clown keep his livestock in!"

"Maybe he's got too large a ranch to fence it all," suggested Marousek.

"Yeah, maybe," muttered Jenkins, as the flatbed up ahead turned onto the even narrower road into Brazel's ranch, a road that was little more than two parallel cow paths, snaking up an incline. Marousek downshifted. The truck cleared its throat and whined up the road. Jenkins checked behind them. He saw the hoist of the tow truck bobbing along in back.

They climbed and rolled down two hills. About a mile in, Marousek pointed. "There's the fence, sir."

But Jenkins wasn't looking at the fence. He was staring at the swath of what looked like pieces of tinfoil, lying to the right. Only it wasn't tinfoil; there was a breeze coming from the northwest. Pieces of tinfoil would be blowing all over the place. This stuff just lay there, shiny and solid and heavy.

By the fence, a man waited. The dun-and-green military vehicles stopped, and Lieutenant Jenkins got out to talk to the man. He wore a hat, a grey uniform, and a badge. A gun peeked up from the holster around slender hips.

"Deputy Horace Brown, sir," the young man said, extending a welcoming hand. "You must be Lieutenant Jenkins."

"Right." Jenkins shook the hand. He was thinking hard, frowning. He knew his weather balloons: the Rawin sonde and the Rawin target. The former consisted of a two-pound instrument package suspended from a helium-filled polyethylene plastic balloon. The latter had the same balloon but it was a radar target of foil and balsa wood. No way either could have contributed those odd pieces of whatever on this field. "What's going on here?" he queried.

"Well, that's why we called the Air Force," said the young man, hitching up his hat and scratching a freckled forehead. "Mebbe you can tell us. Sheriff's up over there past the ridge along with Brazel, the rancher who owns this land. We're gonna have to figure out the best way to get these trucks over there . . . No road, ye see. But for right now, why don't you just come on up an' have a peek."

"Why. What have you got?"

"Like I said, we don't know, but mebbe you folks do. It's some kind of flying device, all right, but it's the damndest thing me or the sheriff or Mr. Brazel ever seen. We figure we don't want to touch it, till you guys come out. Might be radioactive or something." He cleared his throat. "And, sir, there's casualties. Ah mean, fatalities."

"What, it landed on ranch hands?"

"No sir."

"What the hell is it, man!"

Deputy Brown's eyes turned away, a far-off look to them. "Jest come look, huh, Lieutenant? Just come up past this ridge and you'll see *exactly* what I'm talking about."

"I'll bring some of the men with me, then."

Deputy Brown looked over at the others. He drew Jenkins aside and spoke in a low voice.

"Didn't Major Marcel mention to you that this might be, uh, what you guys call 'Top Secret'?"

"What, you mean 'Classified'? Something was said to that effect, but with everybody's noses out of joint about the Russians, that's par for the course." Major Marcel clearly had viewed this whole thing with the same skepticism that Jenkins had, or he would have come himself.

"Well, I would suggest that you come up with me, have yourself a gander, and then you decide exactly who you want over there, and what exactly you want done. And if you've got a radio, I think that maybe your major in charge of that air field's gonna want to have a say too, so be prepared."

The deputy's tone of command and condescension got Jenkins's back up. "I'm in charge here, man. The major has given me full prerogative in this matter."

"Yeah? Okay, it's your ball, I guess. You do what the hell you want. Take my word though, Lieutenant. You take the first look, huh?"

"Very well," said Jenkins. "If that will make you happy, Deputy." He turned to the others, who had gathered around, the curiosity plain on their faces, and ordered them to stay put.

Then he and Deputy Brown made the hike up the ridge.

Out in the open, the air had a charcoal smell to it. Nasty and harsh, like a crash-site back in the Big One, only with no burning petroleum or oil stench.

A few pieces of the tinfoil-shiny material lay twisted amongst the desert brush. Jenkins bent down to touch one, then decided against it. Suddenly, he was feeling a lot less sure of himself. Maybe he should get into one of those radiation suits before he got personal with one of those things. Maybe this whole business was just a little bit more serious than he had anticipated.

Past the top of the ridge, the land immediately dipped into a small valley. Over the crest of another hill, about two miles away, Jenkins could see the tops of the ranch buildings: a long house, some stables, a barn, a silo with the diminutive form of a weather vane topping it.

But that what was below caught his more immediate attention. Down one side of the valley stretched a blacked and burned furrow. Pieces of the shiny wreckage were strewn hither and yon, like some giant robot's dandruff.

Standing by the end of this furrow was a man in bib coveralls and a grey hat. Beside him was a man with a uniform identical to Deputy Brown's, only this guy had a light jacket. The two men were staring down into the depression at something blackened and metallic, maybe forty feet across. Threads of black smoke weaved up, dissipating.

Deputy Brown said nothing, merely gesturing for the lieutenant to continue on down. Sensing them, the sheriff turned around and waved. He looked damned pleased to see a military outfit on its way.

About twenty-five yards farther down, the path skewed in a way that brought them into a position where Jenkins would look directly down at the thing that lay at the end of the blackened furrow. He stopped for a moment and just stared at it.

"Yeah," said Deputy Brown, realizing that the lieutenant could see it now. "That's it all right. Ever see anything like it? The sheriff, he thinks maybe it might belong to you guys, but I don't know." He sidled alongside Jenkins. "Me, I think it's one of them flying saucers that they've been talking about. What the hell it's doing in Chavez County, I don't know. But there it is!"

"Jesus Christ!" whispered the lieutenant reverently. Jenkins could only stare down at the thing and shake his head.

That thing sure as hell wasn't no weather balloon!

That was when Jenkins saw the two blackened bodies thrown to either side of the thing.

Suddenly, he wished things could get boring again.

In his heart of hearts, however, he knew they never would be again.

CHAPTER ONE

The two men blended in with the rest of the patrons at the diner just outside of Hastings, Nebraska. They wore faded old jeans; scuffed Sears work boots, and white undershirts beneath checked flannel shirts. Even though it was spring in the flatlands of the Midwest, the clean, clear air still had a nip of winter in it, and so the men also wore windbreakers as well, which they kept on as they ate their Trucker's Special of orange juice, eggs, and hash browns. They looked as though they could be farmers or ranchers, or in fact truckers. American working men, Joe DrinkBuds straight from some growling Waylon Jennings song. Neither had faces you'd remember. One looked to be in his late forties, with grey touching the sides of his mussed hair—the other maybe in his late twenties with a face that was just a face, an afterthought from the bored genetic pool of America's heartland.

They blended in well with the waning breakfast traffic in that old-fashioned diner with its starched and tired waitresses and the frowning cook straight from *Alice Doesn't Live Here Anymore*. They merged with the faces and clothes and flat speech-patterns because that was what they wanted. If they'd have been on Wall Street, they would have looked like brokers; and if they'd been on Capitol Hill, they would have looked like legislators; and if they'd have been at a Shriner convention, they would have looked like Shriners.

They were here because they had followed the man in the back booth here.

The man named Doctor Everett Scarborough.

Scarborough did not realize it, but they'd been following him all the way from Las Vegas. They did not always have to tail the fugitive—they had other methods of pinpointing Scarborough's location. They even had a pretty good idea where he was headed.

Scarborough kept to back roads, which was smart, and he didn't go into many restaurants, which was smart, too. But his Ford Falcon clearly needed a new fan belt, and he was due for a gas stop, so he'd gotten what he needed in an Exxon station across Route 6 here; then, either from hunger or sheer tiredness, he'd decided to duck into Mort's Good Cookin' Diner for a quick meal and some coffee. Not smart. His face had been all over the newspapers just a couple of days ago, and even though he'd made some effort to disguise himself, things like that flesh-wound scar on his face gave him away. Trouble was, Scarborough was a professional scientist and writer, not a professional fugitive.

The nondescript men in flannel and Old Spice were professionals, though. Professional killers, among other things. But they did not intend to kill Doctor Everett Scarborough . . . not unless they had to.

After they had parked their black Cadillac (a joke among their superiors; nonetheless, with their considerable travel duties, it was a pleasant cruising vehicle) some distance from Scarborough's Ford Falcon in the dusty, unpaved parking lot, they entered Mort's and found him in a back booth of the tatty, yellowed-linoleum place. They'd taken a booth some distance away out of sight of Scarborough's—but positioned so they would know when he left. They did not talk to one another, nor did they read papers. If anyone had taken time to notice, their bodies were remarkably relaxed but their eyes were alert as those of wolves. As they ate their breakfast there were no peculiar eating habits employed, no scraping of toast, no dipping of yolk, no stretching or cracking of knuckles. They just ate. And waited.

Exactly sixteen minutes after they had sat down, two men

entered the diner. The over-madeup waitress with MARGE lettered above a sagged breast showed them to the booth right beside the two men in windbreakers drinking black coffee. Marge handed the men menus. If she had looked at the men in the next booth, she would have seen that they were no longer relaxed. One of the new arrivals was a man in a brown corduroy sport coat and red tie, with a smile like a salesman's. The other was a county police officer.

The older man looked down the aisle toward where Scarborough sat in his booth. The doctor had not seen the cop, and there was no way of warning him. The younger man looked as though he wanted to do something, but the man with the grey in his hair just pinned him to the seat with a glance that said, "Stay put."

The county cop had a beer belly that made him look pregnant; his uniform was a little rumpled, and he smelled of a night's worth of sweat, coffee, and free convenience-store donuts. He ordered scrambled eggs and sausage and a large cold milk, rubbed his eyes, and started talking to the other man about some sort of land deal in which they were both involved. The cop called the man Ted. The man in the tie called the cop Pat. They were just two locals having a breakfast meeting. The men in windbreakers relaxed a little.

Marge brought the sausage and eggs, and the rich spice of the fried meat was heavy in the air for several minutes as Pat the cop gobbled it down. He drank his milk, belched, and then continued talking with Ted the salesman, this time just gossiping about some woman it turned out they had both slept with.

It was at this unfortunate time that Scarborough chose to leave the diner. The man looked tired and lined. Everett Scarborough was fifty years old, but he'd prided himself on the fact that he didn't look much over a grey-touched thirty-five. Now, though, he looked every year of fifty, with maybe a few more thrown in. He walked down the diner aisle with a stiff gait that spoke of aching muscles and maybe a touch of new-found rheumatism. His eyes looked faded and bleary, and his teeth seemed gritted from too much coffee. He didn't notice the officer until he was almost at the cop's table. The two men in windbreakers watched his reaction. They could have been invisible as far as the doctor was concerned, which was good. However, he clearly winced at the sight of the cop, pausing for half-a-heartbeat before forging on, clutching the meal check in his hand, approaching the cash register to pay.

As though sensing something, the cop stopped listening to his companion and looked up at Scarborough when he passed. He twisted around and watched for a moment as the man paid the dark-haired man in short sleeves and glasses behind a counter.

"Shit," said the cop.

"Something the matter?"

"Yeah. I get the feeling I seen that guy's face somewhere before. Like in a picture."

"What, Pat. A Wanted picture."

"Maybe."

"Doesn't look real sinister to me."

"Well, I could be mistaken. But you know, they don't have to look sinister. They could look like Ronald Reagan and be serial killers or somethin'."

Scarborough quickly took his change and left.

"Sure does look like he's in a hurry, doesn't it," commented Ted.

"Yeah, sure does."

"Well, thanks for the meal, Pat." Ted threw down his napkin. "Gotta hit the office. Joyce has got some damned tax forms I gotta go over."

"Good. Pay my salary, Mr. Businessman!"

"Screw your salary. I'm just trying to survive. Gotta run."

Ted got up and left. Pat just stayed in his seat, watching Everett Scarborough get into his Ford Falcon. Making a sudden decision, the uniformed man got up and went to the window. The men in windbreakers watched him, saw his lips moving as he read out the driver's license. Memorizing it.

The policeman went back to the table. He pulled out a ten from a fat wallet and tucked it beneath the bill. Then he stretched and waddled to the back of the diner, where the rest rooms were.

When he was gone, the young man said, "He's not following."

"No, but he's got the license," said the other, tersely. "He's off duty. He's probably just going to radio the state troopers after he relieves himself."

"He didn't seem to recognize Scarborough."

"No, but the troopers might. It will be necessary to detain him. Just for a while, to give Scarborough a headstart."

The younger man nodded. He had a job to do, and he knew how to do it.

While the older man went to pay the bill, the younger man went back to the rest rooms. The men's room was way, way back in the rear of the building, past a low-wattage bulb that flickered on peeling wallpaper. The door was labeled with a generic black Gentlemen metal sign hanging akilter from one screw. The young man opened it and entered the bathroom.

The rest room consisted of a sink, a towel dispenser, a trash can, one urinal, and one stall. The young man had hoped to find the police officer standing at the urinal, peeing. Instead, he immediately saw the two legs by the ceramic basin of the toilet. By the grunts and the smell already permeating the little room, it was clear that the man was defecating.

The young man turned and examined the door. No lock. From the pocket of his jacket, he took out a walletlike case, from which he selected a blue pellet. It was a knockout gas capsule, but to work effectively, it had to be cracked within six inches of its victim's nose. This was why the young man had hoped the man was urinating. He could have simply stepped up behind the cop, snapped the capsule and been done with it.

The young man reached behind his back and drew a gun from its place, tucked in his wide belt. It was a Glock with a built-in silencer. He did not intend to use it—it was just for show.

Fortunately, Pat the policeman had not latched the door of the stall. The young man pulled the door open with his left hand. With his right, he stuck the gun into the door at the man squatting on the toilet.

"Say a word, I'll shoot," he whispered tersely.

The policeman's mouth dropped. He blinked, and brought his hands up slowly into the air, his eyes wide with surprise, but not arguing with the authority of the gun. The man's pants were slung in a ring of disarray around his shoes.

The young man switched the gun to his left hand, and positioned the blue capsule between his thumb and forefinger of his right hand. The gas that the capsule held not only relieved its victim of consciousness for ten minutes to a half hour, it administered a memory blackout for up to an hour previous. Which meant that Officer Pat wouldn't remember seeing Everett Scarborough.

The young man placed the capsule within the proper proximity of the policeman's nose, and squeezed it between his fingers in the manner as he had been trained.

The capsule did not break.

Alarmed, the young man took his eyes from Officer Pat to the capsule. The left-handed gun aim tilted slightly away.

That split-second was all the cop needed. Silently, with a speed that belied his girth, he brought down one thick hand on the fingers holding the capsule, the other on the gun. The ampule was knocked from the hand and rolled away out of reach. The gun coughed, lodging a bullet in cinderblock with a spray of stone. The cop rose up from the ceramic throne like an angry Kodiak bear, one arm holding the gun aside, the other clawing toward the gunman's face.

The fingers reached the young man's temple and gouged. Artificial skin gave way, and the young man's face came half off.

''Jesus!'' said the policeman, stunned, stepping back. ''Jesus *Christ*!''

The young man panicked.

He swung the gun up and shot the policeman twice in the chest and then once between the eyes with remarkable accuracy. Blood sprayed onto metal siding and grey cinderblock. The big man slumped back like a punctured inflatable doll onto the toilet seat. His head fell back, eyes rolling, tongue lolling, as bright red blood dribbled back into his bushy hair.

The young man took a quick deep breath. The policeman was dead, no question about that. An extreme, unnecessary measure, but it worked just as well as the blue capsule.

He put the gun back into place in the small of his back and examined the damage to his face in the smeared mirror above the sink. Carefully, he tucked the skin back into place, sealing it lightly with the proper squeezes at correct pressure points. It was a hurried job, but it would have to do. He and his associate had to leave this place, and quickly.

He found the capsule and put it back in his pocket.

He closed the toilet stall door and then left the bathroom. He was not concerned about fingerprints, because he had none.

The older man was waiting in the car.

As the black Cadillac moved out onto Route 6 the young man applied his breathing exercises, centering himself, releasing his tension and panic in a controlled manner.

"Did it proceed well?" asked his older companion. The car picked up speed, hurtling beneath the big blue of the midwestern sky broken only by puffy cumulus.

"No," said the young man. "I had to kill him."

"Kill him! We kill only when there is no other choice!" The older man gripped the steering wheel, his knuckles growing white.

"He saw my face. My *real* face."

"Unfortunate. We'll have to take the time to make a complete identity and registration change. We can't take any chances that they might remember us, or the car."

"Yes," said the younger man. "I'm sorry."

He told his companion of the trouble with the knockout capsule.

Afterward there was silence for a long moment.

"Well, we know where Scarborough is going. We can afford to lose him for a while. Let us just hope that he doesn't need his guardian devils." A humorless smile touched the older man's lips.

The younger man, however, did not smile at all.

He was too busy staring out the back window, watching for any pursuit, doubtlessly wishing that this, his first assignment, was not so terrifyingly and crushingly important.

CHAPTER TWO

She was in *control.*

As the last hypnotic commands issued from her mouth, she felt very much in charge of the situation, very much at peace. The room seemed to buzz with an intense respect, a great awe at her power. Everything was in its place, the blinds drawn just so, the illumination at the exact wattage she'd requested. Earlier, before the session, there had been a fly in the hallway. She knew it was there; she'd heard its fitful buzzing as she'd prepared for her patient. Within minutes, two agents were out there with swatters and sprays, and pretty soon that offending fly, that dreadful insect that might somehow infiltrate her session, was terminated and removed.

"You are feeling very relaxed now," said the doctor to her patient. "Very calm and relaxed."

"Yes," said the man in the chair. "Yes."

The doctor got up from behind the desk, quietly opened up the cabinet behind her. Inside were some of the supplies she would need for the session. Now that the patient was under, it was safe to reveal them; he would notice only the things that were called to his attention.

Then she went back to her desk and pressed a button.

In another part of the building, the suitable people were notified that the doctor's patient had succumbed to deep, deep hypnosis. They would be ready within minutes for her next signal. She opened up a drawer and pulled out a sheaf of papers. Her script.

"Let us go back to the little house on the island, to the night of February 28, 1986," she said to the patient. "Let us go back to the night when the aliens first actually placed you in their flying saucer.''

The man in the chair nodded. His face was relaxed. Though it was not an old face, already there were pronounced wrinkles about the eyes, the mouth, and the forehead, due to the extreme tension he had been through these past years.

''Yes, yes,'' he said, and his voice was cultured and precise, with excellent diction. ''*That* night.'' He rolled his shoulders as though acclimating himself to his new setting. ''I have finished my work late today, and I am relaxing with a cup of cocoa on the veranda. I am alone, and yet I am not at peace. Something is deeply disturbing me. Something that vibrates through the air. As I am sitting in the chair, I see a light flashing off in the clouds over the ocean. An airplane? The light shimmers through the low clouds. It moves too fast, too jaggedly, to be a plane. It disappears. This unsettles me deeply, and yet I am tired. Very tired. It has been a long day, and I want to sleep. I lock the doors, and I bolt the windows and perform the other rituals that have been my compulsion for the last week; why, I don't know. I go to sleep as soon as my head hits the pillow.''

''Yes,'' said the doctor, putting on her gloves, beginning to lay out the necessary instruments from the cabinet upon her desk. ''Yes, and what then?''

A spastic tic quirked the whole side of the patient's face.

''What happens then?'' continued the doctor. ''Do you sleep through the whole night?''

''I wake up—in a cold sweat . . . And I see . . . I see at the other side of the room . . .''

The tic again. Damn, thought the doctor. The dosage was too low. He needed more and more each time! Quickly, she picked up a fresh hypo, tore off the sanitary wrapping. It had already been filled with the drug solution, waiting for her on the tray, just in case.

The patient's tic had changed into a spasm that was contorting his entire face. His feet began to twitch, his torso began to shake, his tongue began to protrude from his mouth. The doctor imagined the storm that must be raging now inside the man's brain. All the signs of an epileptic fit.

But of course it was not epilepsy.

She grabbed the man's hand and pulled it out. The needle

stabbed down, buried itself between the soft juncture on the opposite side of the elbow, and she pushed the plunger.

A split-second later, the man relaxed, lolling back in his chair.

The doctor sighed. She removed the needle, ignoring the little trickle of blood that followed it. She placed the hypodermic back onto its tray, and then turned to the sagging patient. She swung an open palm across his face. The slap resounded through the while examination room, and the man's eyes fluttered open. "Oh! Oh!" he gasped. "Yes. Hurt me! Hurt me, I *deserve* it."

"Just a little love tap to adjust your senses," she said in a professional monotone. And, she thought, We're going to have to make run another biochem analysis on him, soon. Serotonin might be a little sparse. "Are you back in your room now?"

"Yes." His eyes were open now, wide. "Yes, I'm back in the room."

She went to the desk, hit the buzzer to summon her "helpers." "And what do you see?"

The door opened. Three men walked in. They wore masks and silver suits, and carried rods that glowed a pale phosphorescent green. The masks were oval-shaped with bright almond eyes and narrow V-shaped chins.

The doctor held up her hand for them to stay where they were.

"Perhaps you should look again," she instructed. "Perhaps you should look behind you."

Slowly, the man opened his eyes. "No. No, I don't want to."

"Look behind you. You have to face up to your memories. You have to delve deep into them, to discover the truth about what happened to you."

Slowly, the man turned. When he saw the men in the masks and the silver suits behind him, he whimpered. "No," he said in a soft voice. "No."

The doctor held up her hand and beckoned to the three men. They started forward, raising their rods, which began to flash softly.

"What do you see in the room!"

"They're here again! The Visitors! Three of them! They have their light-rods." The patient gasped, drew in a deep breath. "Stay away! Don't come any closer! Don't touch me!" The man moaned and cringed back, but he did not move from the chair. It was as though he was stuck there by some invisible adhesive force.

The three agents in masks and suits stepped forward, slowly in an awkward, stylized shuffle, like clones of Charlie Chaplin's Tramp in slow-motion. The frontmost held out the glowing rod as though he were some magician, about to pronounce a benediction.

"What's happening now?" the doctor demanded.

"They're coming . . . coming . . . Omigod . . . don't . . ."

The lighted end of the rod—the effect produced by a simple battery and bulb arrangement—touched the man softly between his eyes.

The effect was immediate. The man crumpled as though someone had clicked a lever in his spine, causing him to fold in on himself. Despite the injection, the man's hands were still trembling, and a tic constricted the left side of his face, as though an invisible hook was yanking it up the skin, then letting it go, yanking up the skin, letting it go.

"On the examination table," said the woman to the agents.

One of them flipped up his mask, and looked down with concern at the patient. "Look's like he's having some sort of preliminary fit, Doctor. Maybe we'd better just skip this session!"

The doctor glared at the man. She was an attractive woman in her mid-thirties with high cheekbones and long beautiful hair, but the hair was drawn back and tied in a utilitarian bun, making the angles of her face severe. Her lips were thin, and her eyes were dark; and the combination of a frown on those lips and glare from those eyes were rumored in the Company to have been utilized from time to time as an Extreme-Prejudice measure on double agents.

"I will brook no insubordination, man! *I* am in charge of this operation. Get the patient off of the chair and put him on the table. Now!"

The agent slipped the mask back over his face. He and his fellows grabbed the patient by his arms and legs, carried him over to the vinyl-cushioned table, and placed him on the sterile paper covering.

The doctor rolled up her instrument cart, then took her script from the desk and turned several pages.

"We're going to proceed with the directives on Page Eight

dash B today, gentlemen. So if you'd be so kind as to remove the subject's clothing."

The man's hands had begun quivering as though he were a victim of palsy.

"Eight-B . . . Isn't that a bit *extreme* for someone clearly in his delicate shape."

She could not read the expressions on the agents' faces, nor did she particularly care what they were. "I'm the doctor. Let me be the judge of that."

They removed the clothing, leaving the patient totally nude upon the table.

"Restraint," said the doctor in a monotone.

Two leather straps were pulled from the other side of the table, and securely buckled into place across the patient's chest and abdomen.

"Examination aids."

From the bottom of the table, two gynecological set-ups were protracted, angled, and screwed tight. The patient's legs were then strapped securely into the stirrups.

The man's body gave off the smell of deodorant and sweat. His hands were still shaking, and his teeth chattered like broken castanets.

The doctor tested each of the restraints. "Fine. Leave now. I'll summon you when I need you."

"He'll be all right?" asked one.

"He'll be just fine."

The men left the room obediently.

The doctor went back to the cabinet and pulled out her own mask. This was a professionally designed mask much the same as the others, but much more realistic. It was made of latex and it fitted easily over her head, with holes for her eyes, nose, and mouth. Also from the cabinet, the doctor took up one of the wand devices. She turned this on, and examined the scars she had placed on the patient's torso over two years ago, after all this had begun with him. There was the triangular-shaped scar on his left forearm; the lopsided figure-eight sign on his right shoulder. A little more delineation next session, perhaps. Best to keep him on his toes about that. The scars kept his mind constantly on the phenomenon, and that's exactly where she wanted it to be.

She tapped him twice on the forehead, and gave the command to rouse from unconsciousness and back into his hypnotic state.

"Now what is happening?"

She leaned over and whispered the word *Pequod* into his ear, unleasing the pre-established reality construct.

"I'm in the saucer," he said. "I'm in the white room. The examination room. And standing before me . . . Oh, Jesus! It's here again! It's Eve."

"Eve is only your word for her, remember."

The patient struggled. "They've got me tied down. Oh, God, what are they going to do to me *this* time?"

The doctor smiled grimly underneath the mask.

She took the lit rod and let it drift down the man's chest, his abdomen, down to the pubic nest, where it briefly touched the man's flaccid penis. The man squirmed and struggled, his palsied fingers wriggling like electrocuted worms. The rod slipped down further into the cleft of his buttocks, where it lingered for long moments.

"No! No!" cried the man. "Not that again! No, stop it! Who are you? What are you doing?"

The doctor pulled the rod away. She sat it down onto the tray. No, that wasn't necessary for today's session. He didn't need that kind of reinforcement this time.

"Who are we?" the doctor said, using her high-pitched Eve voice. "You know that by now. We are visitors from the planet Draco. We are performing experiments upon you human beings. We need fertile sperm-plasm, Earth-one. We need to replenish our species with a stronger genotype, a hybrid species."

So saying, the doctor unbuttoned her blouse and her skirt. She stepped out of her shoes and her underwear, and then unhooked her bra, loosing her large-nippled breasts.

"Oh, please. Please, don't you understand. This hurts me. Don't do it *again*!"

The doctor picked up the stainless-steel milking device from the tray. She considered lubrication—but no. There could be no pleasure in this session. Not in this delicate part of the treatment.

The patient began to screech pitifully.

"Please help us to help you stop screaming," said the doctor as she advanced with her device toward her patient.

Yes, she thought. This was her favorite part.

* * *

That afternoon, the superior, who was also in charge of Operation White Book, came to Dr. Julia Cunningham in her office.

"I'm glad you're here," she told him as he accepted the cold mineral water she offered him. "We have some serious things to go over."

"Yes, I'm sorry about the delay. We were scheduled for three days ago but what with all that's happened . . ."

Brian Richards had the chiselled good looks of a character actor usually cast in Westerns or hired to sell Marlboro cigarettes. He was slender and looked good in his expensively tailored but unassuming grey business suit and muted red tie, with a white handkerchief peeking up from the top of his pocket. His dark hair was peppered with grey, but rather than making him look old, the grey, along with the network of undoctored wrinkles over his face, gave him the look of steely and unquestioned authority. Now, as he sat in her chair calmly sipping the water she'd given him, he reminded Julia Cunningham of Dr. Preston Cunningham, her illustrious father.

And she hated him just as much.

"Yes. I've gotten the reports," she said blithely, sitting stiffly in her chair, like a statue of cold metal. "But I've been busy with my other White Book projects. Would you care to give me a quick, relevant briefing?"

He smiled coolly and put his water down. He looked at her so piercingly with those flinty eyes with twin sparks of humor dancing in their darkness, that she had to divert her own eyes. There was a twisted smile on his face that demanded challenge.

"What are you looking at?" she asked defensively.

"I'm just wondering about that saying, 'Butter wouldn't melt in her mouth.' Care to experiment, Doctor?"

"What are you talking about, man?"

"Dr. Cunningham, you must know that Woodrow Justine was killed in our little operation last week."

"Yes. I was given that information. So?"

"So? Doctor, you of all people should have some kind of reaction to that news. Woodrow was your patient. You worked with him for years." That wry smile again, damn his eyes! "You more than anyone else made him everything he was!"

"Richards, you're the last person I have to remind that Justine was a hired killer. I merely adjusted his psychoses sufficiently to within malleable parameters, that the Editors could make use of his . . . considerable talents. Need I remind you, Mr. Richards, that controlling killers is not my main function here at White Book. Now, perhaps you'd care to proceed with the more important matters of our operation."

The smile faded. Clouds filled the man's face. He licked his lips, turned away for a moment, and when he finally turned back to her, there was lightning of anger in his eyes. "You *staked* your *professional reputation* on Justine's performance, Doctor!" he said through gritted teeth. "You said you'd swept the demons right out of his head. And then he can't deal with a leftist-leaning, nonviolent chump like Everett Scarborough! He goes over the goddamned Hoover Dam for chrissakes! And we still can't find the body!"

She stared at him for a cool moment, and then said, "Need I remind you, Mr. Richards, that there is no record of what occurred on the dam. It was you who endorsed Justine's preference for solo operations. His backup was engaged—elsewhere. Quite simply, we don't know what happened. So, you cannot, in good faith, call in that particular card." She stood up and leaned on her hands, putting her face a little closer to his, challenging his space. "Besides, Mr. Richards why *should* you? White Book is at a critical stage. You, more than anyone else, know that I *am* White Book, Mr. Richards. Why would you wish me gone at this point, anyway? Why are you acting in this aggressive manner toward me?"

Richards's frown turned into a nasty grin. He sat back in a chair and crossed his legs carefully. "Mind if I smoke, Doctor?"

"Yes. You know my feelings on the subject." She tapped the sign on her desk that read No Smoking Please right by the international symbol of a red-halved circle over a burning cigarette.

Richards pulled a pack of Benson and Hedges from his jacket pocket, mouthed the words "Fuck you, bitch" through a smile, and pulled a cigarette out. "You know, Julia, I can remember when we used to be a lot nicer to each other. Gee—way back, when I saved you from Stanford U. There you were, a waif in graduate school, bucking the system with your ideas, staring at a bunch of hostile teachers and psychiatric chumps. About ready to knuckle under so you could get that old diploma, as I recall. Then *we* stepped in and gave you all you dreamed for and more . . . A quick degree from Johns Hopkins U. Some lab work at NIH, consulting at NIMH, a few experiments in the field . . . And suddenly, the sweet little genius who Daddy'd cut off without a cent was pulling down six figures, with perks coming out of her pretty rear-end.'' He lit the cigarette, puffed a moment, and glanced at her with an appraising look.

''You mean, those times I let you sleep with me, Richards,'' she said, contempt showing in her eyes and her words. ''Is that what you're talking about?''

Richards shook his head. He looked down at the cigarette, licked his fingers and put it out, tossing it into the waste bin by the desk. ''No, Julia. No, that's not what I'm talking about—that was a mistake, on both our parts. I'm sorry.'' He closed his eyes. ''Pressure, Doctor. Pressure. That was long ago and far away, and I apologize that you thought I was bringing it up. I apologize for my anger, Julia.''

She relaxed despite herself. Damn him! He always did this to her, the bastard! Pushing her buttons so that up went the Shell—and then making her feel awkward, putting her off guard.

''It's just a rough time, I guess. And with Scarborough not a pawn anymore, but a wild card . . . Well, I guess I'm a bit on edge.''

''Would you like me to prescribe something for you, Mr. Richards,'' she said tightly, still in control despite herself.

''No. No, thanks, Julia. Occupational hazard. Just keep a nice pair of lungs and a new liver for me on ice, should I need them.''

''I can't guarantee anything on that, Richards. You shouldn't smoke; you shouldn't drink. Period. Doctor's orders. Let me do an analysis, mix you up a pill suited just for your neurochemistry. You'll think clearly but calmly. It's very simple, really.''

''You'd like that, wouldn't you, Julia?'' His mouth was smiling but those cold eyes weren't. ''You'd like to be my pusher, wouldn't you? More control over the old man, huh?''

''That's not . . .''

"Cat and mouse, Julia. And why not? Part of the fun. Part of the games. We're a real pair, aren't we?"

A sudden involuntary flash of recollection suddenly shot through her mind. She and Richards lying naked in her Silver Spring apartment bed, lit only by the light of her aquarium.

She clamped down a hard shutter on that memory, poured a glass of mineral water and sipped it, as though to wash it and *all* her unwanted feelings away. Dammit, she was the one who needed a new psycho-neural prescription! Physician, heal thyself!

"I don't know what you're talking about, Richards," she said. "I think it best, in light of the situation, to pare our conversation down to relevant points." She leaned over and spoke tersely, with perfect diction and exaggerated intonation. "In short, let's keep it professional."

Richards nodded agreement. "Yes, and we are professional, aren't we, Doctor? So then, to the specifics of our respective professions. I trust that you have your report on the progress of White Book?"

The folder was to her immediate left, appropriately colored a bright lambs-wool white. She handed it to him. He opened it and proceeded to peruse the neatly typed pages on the spot.

Project White Book, thought Dr. Julia Cunningham, watching Richards read, knowing that the information expertly sketched out in her excellent prose would please her superior. How much this project had changed over the years since its inception back in the fifties.

No, not changed.

Mutated.

That was the correct term for it, really. The verb, anyway. And the noun? The synonym?

Well, surely *that* had to be that wonderfully coined government term, that bit of bureaucratise that fit the bill so perfectly, it immediately became an integral part of that very nonspecific language.

Disinformation.

That's what the CIA started calling what the KGB and the Russian media were doing back in the seventies. Only, of course, they'd all been doing it to some degree, ever since the cold war began. Disinformation. Turn the available intelligence material available into shit. Gobbledegook.

Project White Book. The program behind Project Blue Book.

When some the Air Force officials had spoken out later on just *how little* Project Blue Book had done to investigate the entire UFO phenomenon, they couldn't have been more wrong. Project Blue Book had been a vital cover, the token effort by the CIA tools in the Air Force to placate the calls for official investigation for lights in the skies. But the *true* operations of importance, instigated very early on, were White Book and, of course, Black Book. Cunningham knew very little about Black Book; just enough to keep her department in White Book up to date. Besides, Black Book was pretty exclusively Richards's department—and it was there that the bastard was an underling. He was Editor-in-Chief, sure . . . But over him in Black Book were the Publishers. Maybe that explained his moodiness . . . yes . . . Julia Cunningham had to chuckle to herself. That was it! The Publishers were nailing his balls to the wall, and they could do it, too. They didn't bother her much, and she was grateful for that. Richards didn't scare her, the CIA didn't spook her, not even the psychotic crazy Woodrow Justine scared her. But the Publishers . . . She never admitted it to anybody, but they scared the shit out of her. They'd be the last people that she'd cross.

They were the ones, of course, to have invented the White and the Black. Oh, Richards claimed that the CIA had started it on some level or another. Top secret, hush-hush, and all that. But from what Cunningham knew and *intuited* about the Publishers, they were somewhere down in the dirt from the beginning.

Project White Book, of course, was the machine, surreptitious and circumspect, that had actually *created* many of those sightings in the fifties and the sixties. Oh, natural phenomena and human imagination made most of them—you planted one suggestion in one Arkansas hick's mind, and suddenly, the whole town was seeing saucers over Farmer Jones's silo. Plant enough sightings, and pretty soon you had some kind of control of the situation. That had been the theory back in the fifties, and it had worked for a while. But then, the business had changed, the mythology had grown . . . Betty and Barney Hill had been abducted in the early sixties, John Fuller had written *The Interrupted Journey* back in 1966—and suddenly, direct extraterrestrial contact was the state of the art. By the time Travis Walton got picked up for his *Enquirer* prize-winning joyride in the mid-seventies (not a White Book project, but rather what the Company Branch dubbed "a grass-roots" hoax), the wave of UFO-ological future was written on the walls. It was in 1977 that Brian Richards, then just an Editor himself, had found that promising student at Stanford. By 1980, she and her methods were put into full service. When Richards was elevated in '82, she got her full Editorship, and reported directly to him.

The theory wasn't hers. That had come down from the Publishers. But the methods were, including the science, the hard work, the gains, the failures—the outrageous successes that were the cornerstone to the entire program. Richards had explained it to her in blunt, simple terms. "I'm sure you're aware of the office-full-of-women phenomenon. The pheromone leader has her period and it keys into the other women. You have a bunch of women in the same office, in close quarters, for a while, and pretty soon the janitor knows just when to stock the WC tampon dispenser every month. This is kind of what we're doing, Julia."

They'd been lovers then, and he had chosen that time, lying in bed twined in satisfied sheets, to tell her. "Back in the fifties and sixties, it was scattershot, but eventually we picked the right 'leader.' Only it wasn't pheromones, it was kind of culturally psychic—a cosmic fashion-leader. You found him or her, you showed him a flying saucer buzzing his rooftop, you gave him the right suggestions—bingo, not only do you have a convert, but you have a 'leader.' And the magic is, he or she doesn't even have to write about it or be on TV; the psychic 'scent' just sort of wavers through the *zeitgeist*, the collective unconscious—gets pounced on and 'stepped up' by the saucer freaks, the true believers, maybe even changes the very fabric of reality. And suddenly, there's a new public awareness, controlled by us.

"We find the 'leaders.' We get them to you, give you the time you need. The complicated part is up to you. You give them a wild trip. You program them in such a way that they believe they've been kidnapped, then prodded and poked by aliens. We manufacture a continuingly bizarre and escalating scenario behind it, but mind we don't get too consistent or make too much sense—that would rouse too much suspicion in scientific, government, and academic circles. A few of these, properly placed, every year. Voila! The bread falls upon the waters and disseminates! The pebbles become snowballs, which become an untraceable avalanche! Instant phenomena!"

The whole idea had been so exciting to her—such control, such an opportunity to experiment, such a wealth of knowledge of the human neuro-chemical system to be gained—that at first she didn't ask the obvious questions. But eventually, the curiosity had gotten the best of her.

Why?

What was the reason behind all this?

"Oh, you'll find out soon enough," Richards had told her. "Just know that this is for your country, that there are forces we are fighting *against* in this vague but effective manner."

But by the time she'd found out the entire truth, she was a full Editor, and far too involved, emotionally, professionally, and otherwise, to back out.

Besides, she couldn't back out now if she wanted to. Project White Book was her life, the things she was learning, accomplishing, were absolutely *incredible*. The sense of control over other human beings, over her own destiny, gave her an incredible sense of *power*, far better than drugs because it was *real*, because it *lasted*.

Anyway, if she left, they'd kill her.

"Good," he said. "Very good, dear Doctor," said Richards, pulling up his briefcase and neatly tucking the folder and its contents inside. "Things are coming along well here in the wild and wooly west." He looked around the office, smiling. "Yes, this branch always seemed to bear the most fruit for some reason. You think it's something mystical, Doctor?"

"I think it's fairly isolated, Richards. Easier to operate, easier to keep secret...more room to move around in." She tapped her coffee cup. "Yes, and all the other military operations going on—well, the locals are used to seeing helicopters swooping through the night and mysterious, draped military vehicles roll between the mesas and arroyos."

"My dear, how unromantic. You'd never make a good saucer nut." He got up from his chair. "Now, for my tour. And to attend to the reason for this little side trip. Your report sounds most promising. Is full implementation imminent?''

She nodded. ''Our subjects are responding quite well to the conditioning. The introduction of the Millennium factor was brilliant. I think we've got ourselves what amounts to a new religion on our hands. It's very exciting to be a part of its birth. It's a belief that has its beauties and elegances . . . as well as its stark blacknesses . . .'' She stopped herself, suddenly realizing that she was expressing enthusiasm, revealing emotion. This would not do. ''Scarborough,'' she said, changing course. ''He's very dangerous to us now, you know. Very dangerous to White Book . . . and I imagine that he is dangerous to Black Book as well.''

She thought about Scarborough and a kind of darkness trickled through her mind. Anger. And other feelings she immediately repressed. The Shell again. Up went the Shell. Richards mustn't see . . . He mustn't perceive her true interior, as riddled with faults as Southern California.

''Don't worry, Doctor. Thanks to our media connections, the public is starting an about-face in its perceptions of Everett Scarborough. No longer is he the champion of logic and reason, the clever and articulate skeptic—he's a crazed murderer now . . . A charming psychopath who has fooled the public for years. If he comes anywhere near the real truth and tries to express himself . . . Why, by then, the harm will have been fully done to his respectability. He will be perceived as a paranoid of the very first water . . . A hapless victim of reading far too many Robert Ludlum novels.'' Richards smiled and pulled out a thick book. ''By the way, I just finished the latest Ludlum on the plane. Care to borrow my copy?''

She shook her head. ''I don't read fiction, Mr. Richards, you know that.'' She returned to the previous subject. ''But Scarborough . . . You *are* trying to catch him, aren't you?''

''Why yes, of course.''

''And kill him?''

''We can't allow him to run about free now, can we? Not with what he seems to know now.''

''A thought, Richards. Kill him by all means, if it's necessary. I understand your position. But if you manage to merely *capture* him, and you wish to question him . . . before you terminate . . .''

She paused, groping for the appropriate words.

''You want him, don't you, Doctor,'' whispered Richards, a hint of a vicarious thrill in his voice.

''Yes,'' said Dr. Julia Cunningham. ''Let me have the great skeptic for awhile.''

Despite herself, a small smile crept over her unglossed lips.

CHAPTER THREE

The one thing nice about getting assigned to Nowheresville, U.S.A., was that the houses were a hell of a lot cheaper than the bigger metro areas.

Marsha Manning had bought her pleasant Cape Cod–style home, complete with a precise acre of tree-spotted and lawn-covered land, with the help of a VA mortgage and a small inheritance from her grandfather for the down payment. She was getting close to two years' worth of equity, which was nice; but two years also meant reassignment time. She wouldn't lose a drop of money in the move, and maybe she'd get an even better deal wherever she was going—and just maybe she'd get an apartment there and rent her house out. Whatever happened, she'd have to leave the house, and she'd certainly miss it, no question about that.

Marsha Manning was thinking this as she turned the wheel and drove her blue Hyundai Excel off Kepler Drive and into the macadam apron of her driveway.

A late spring afternoon sat gently upon the slate shingles, the red brick, tousling the leafy and grassy hair of the trees and the ground with its breeze like a fond parent. The smell of fresh beech leaves and tulip blossoms from her garden rolled out to meet her. A bunch of kids were skate boarding by the curb up the road. It was a family-oriented suburb, a friendly, safe place. She knew most of the neighbors, and they were always trying to fix her up with some cousin twice-removed, or the nice accountant that lives down on Elkins by himself, after his wife left him. What was a young single woman doing, living here? they probably asked themselves after she left their coffee clatches or Tupperware parties. Goodness, she must be awfully lonely in that big house!

Marsha turned the ignition off, put the emergency brake on and hauled out her brown bag of groceries.

Actually, what she was doing here was cooling her heels after a long relationship. Actually, she just wanted to be by herself. Oh, she'd go out with some of the guys Jill Perkins and Harriet Durk set her up with. It wasn't like she'd decided to be anti-social or anything. She just didn't want to get involved emotionally, and so when she got assigned to this area, she'd bought herself a nice comfy house and started working on some personal psychological housecleaning. Out with the dust of dependency! Strip that yellow-wax buildup of a decayed love! In with some bright new furniture of reading! sports activities! subliminal self-improvement tapes! Paint these drab walls a new Marsha Manning hot pink, and start to feel good, like a competent woman of the nineties should!

And all this was going very, very well indeed, until she'd met Doctor Everett Scarborough.

She got out her key chain (complete with a miniature circuit-board the guys at computer training had given her when she'd aced them all in finals—haha) and unlocked the door. Who would have thought that an eminent scientist, author, and self-proclaimed UFO debunker on the wrong side of fifty would have sent her neat emotional interior-decorating into disarray? To say nothing of putting her professional career teetering on the brink...Everett Scarborough. Well, she thought, switching on the light, she'd done an incredible song-and-dance to her superiors, and at this point it didn't look as though she was going to get court-martialed (if Colonel Walter Dolan had more power, she would have been). Thank God she'd taken the time to do everything as by-the-book as possible when she'd yanked Scarborough out of that Air Force brig. And thank God the general she'd gotten out of bed to sign the papers was someone who liked her and owed her a favor. Otherwise . . .

Well, she just didn't like to think about it. The Air Force was her career, and while she wasn't exactly on a honeymoon here, it was a commitment, a relationship, a—

She set the bag of groceries on top of the kitchen table and looked around, frowning.

Something was wrong here.

She could sense it.

Everything looked in its place, all right. The zinc of the sink was fresh and sparkling, the coffee mugs with their funny sayings and Far Side cartoons hung just so in their placements, and the cabinet doors were all closed snugly.

All the same, she had this sensation on the base of her spine. Not fear precisely, just a kind of catlike wariness.

Was there a burglar in the house?

Her internal alarm was still ringing.

As it happened, there were a couple of local police—Jim Nichols and Rob Prosky—who'd taken a fancy to her. She'd gone out with Jim once or twice and she'd have gone out with Rob, only he was married. So she had a good relationship with the local police. "Call us anytime, Marsh. You've got our number—911. You even get a *whiff* of a problem, babe—we'll be painting the streets with rubber to get to your house. Even if you just need someone to have a coffee and a donut with!"

She'd never called them before. She didn't want to cultivate an impression of helplessness with anyone. That just wasn't the Manning style. Besides, there'd never been a good reason. Not even when Pressy, the neighbor's cat, had gone up on her roof. She'd borrowed a ladder and gotten the creature down herself.

There was something wrong here, though, and after what had happened to Mac MacKenzie out in Iowa—well, Marsha Manning was a lady who knew where pride should end and pragmatism should begin.

She wasn't exactly a phone person, so she kept only two in the house: one in her study (so she could hook it up to her modem for her IBM PC), the other in her bedroom. The bedroom was closest, since it joined the kitchen through a short hallway, so she went there.

Maybe she was being silly, she thought as she turned on the

light. Better, though, to be silly and cautious, then to end up violated or dead.

The princess phone was by the bed. She went to it, picked it up, and started to dial.

Nine.

One—

The man must have been hiding in the closet. He came up behind her and put one hand over her mouth, the other around her waist.

"Don't move," he said. "I'm not going to hurt you. Is anybody out there with—"

Even before basic training, Marsha had taken some martial arts classes. Her response was automatic and reflexive; she simply used her body as a fulcrum, pushing her butt back into the man's legs and hauling him over her shoulder. He landed in a surprised sprawl on the floor.

"*Hai*!" she cried, lifting her hand and making a weapon of it. She was about the bring it down on the intruder when his voice registered in her head. She stopped and stared at the upside-down face. That registered as well.

"Scarborough!" she said.

Lying by her feet was a face recognizable as Dr. Everett Scarborough's, but just barely.

"Christ, Marsha!" he moaned. "I didn't know you were a Master of the Far East, or I would have used a stage whisper from the closet." He shook his head, rolled over, and staggered shakily to a kneeling position. "I didn't see . . . did you come in with anybody?"

"No. I'm alone. God, Everett! God! You have my phone number. You could have called!"

He shook his head. "No time. Christ, the house may even be watched. But I had to take the chance. I need your help. Really badly."

He got up and sat on the edge of the bed. He was Everett Scarborough all right, but he wasn't the Scarborough she'd first met at the Iowa farmhouse of Mac MacKenzie. He had about a week's worth of beard growing for one thing (on a formly impeccably shaven face). His hair was shaggy and unkept where it had once been well-groomed; he wore dirty jeans, a torn flannel shirt, and a blue windbreaker, where his previous attire had generally included ties even at his most casual. His eyes had bags under them; he generally looked pretty haggard.

She was speechless.

"I got in through an open window today. I was so tired I fell asleep on the bed. When I heard you come in, I didn't know if you were alone. Sorry about the announcement of my presence. I wasn't thinking straight."

"Ev . . . Ev . . . What are you doing *here*?"

He looked down at his scuffed walking shoes and for the first time ever Marsha saw him looking almost defeated—and definitely very vulnerable. "You said . . . back at National Airport. You said that if I needed to, I could come here. You gave me the address."

"Of course you're welcome here," she said, a bit at a loss. "I didn't mean it that way, really . . . I mean, what's happened . . . I read about that government agent falling off Hoover Dam . . . About your disappearance . . . I honestly didn't think you'd turn up here."

Scarborough nodded. "Yeah. They didn't release the full story. Actually, they want my ass. Bad. They just don't know how to couch the words. Hey, you think I could have some coffee? My head feels like it's filled with cobwebs."

"Sure. Think you can make it into the kitchen?"

"As soon as my spine and a couple ribs mend."

She laughed despite herself. "You live and you learn, Doctor. Lie on the bed for a while if you want. I'll go in and grind some Colombian and plug in the coffee machine."

"Thanks. But I think I can struggle out."

He moved creakily, but he did make it out to the kitchen, where Marsha proceeded to start up the coffee.

"You want something to eat?"

"Yes. I'm pretty hungry."

"How did you get here from Nevada, anyway?"

Scarborough put his head in his hands and sighed. "Maybe you'd better get some coffee into me first, kiddo, and then maybe a quick shower. I feel like something the cat dragged in and then buried in the kitty litter box."

She fed him a cup of coffee and then showed him where the bathroom was and gave him a fresh towel.

When she heard the thrum of the water hitting the tub, she

went out and poured herself a cup of coffee, added some Equal and then some 2% lowfat milk.

It was only then that Marsha Manning permitted herself a slow, slight smile.

Well, so Everett Scarborough had come into her life again streaming clouds of strangeness and stubbornness, mystery and adventure!

She was happy to see the guy again.

Very happy.

CHAPTER FOUR

Marsha Manning fixed him a hearty meal of eggs and bacon, homefries, orange juice, whole wheat toast, jam and anything else he cared for that she had in her icebox.

It was delicious.

On the road, he hadn't eaten much; hadn't been able to keep much down. Besides, he'd kept to the backroads, away from the high density civilization along interstates, shying away from well-trafficked truck stops and McDonald's, keeping to small town Mom and Pop stores where he purchased hasty sandwiches. He ate these gastronomic wonders in his car, keeping the beat-up Spark car radio on any local all-news station he could find; or failing to find such, enduring the groans of country-western or the moans of Top Forty to catch any breaking news bulletins about him, about UFOs, about the further work of the outlaw government organization that was out to get him.

He got nothing new.

There weren't even any real follow-ups in the newspapers. Just one big splash in the next two days, and then zip. Or anyway, nothing that he found in the papers he could get ahold of in 7-Elevens or from machines. Coincidentally, he had seen the sensational report just the other day on *A Current Affair*, when, totally exhausted he couldn't take another night in the backseat and hazarded a stay in a ramshackle old neon-sputtering motel in on the outskirts of Muncie, Indiana. Of course the ironic thing about that show was that most of it had been in the can for weeks; he'd done special interviews for that show to go along with his new book, *Above Us Only Sky.* All the sensation-sharks had to do was to change the slant of the piece, toss in some fevered commentary, trot out some stern-faced troopers and officials, show the bloodstains on Hoover Dam and Ka-zam! Instant trash-TV! Scarborough felt as though he were watching some other guy mouth those tart zingers about the credulous, the unwashed, and the uneducated. Now, after all that had happened, he was a different man. Exactly who, though, he had no idea.

God, it was *so good* to be in a friendly place, to smell friendly cooking smells, to see Marsha's little magnet knick-knacks on the refrigerator, the homey quilted pot holders hanging near the stove. He longed for his own home in Bethesda desperately, that little oasis he'd built himself, tucking his heart and soul away from life's brickbats; it had been especially important to him after his wife's death, since it held most of the memories of her still decorating the inside. It was his big, semi-expensive nest, housing his library, his stereo, his record collection, his desk and word-processor—all the lovely things that insulated him against the outer chaos, the cold and the dark of life. Warmth, reason, intelligence, clarity; these were the things his house held for him. But now, of course, Scarborough didn't know *when* he'd see it again, if ever, and so he took great gratification in the little details of Marsha's house. The old grandfather clock, the antique chair in the dining room, the sprawl of broken-spine paperbacks and winged magazines over in the living room. It gave him a feeling of home, but even more it gave him more hints into this woman's life and personality.

"Hmmm," Marsha said, watching him as he tucked away the eggs and crunched the bacon as they made the small chitchat he insisted upon, asking her this and that about the house and her life, talking about *anything* but UFOs and the dilemma he was in. "Guess I'd better make you some more."

She did, and he dug in heartily.

Under normal circumstances, Everett Scarborough might have objected to the high cholesterol content of the meal. At his last yearly checkup, the HCL had clocked in at a touch too high, and he'd been cutting down on fatty stuff lately. However, since he was hungry as a horse and didn't know if he'd live past the next week anyway, he didn't make any objections.

He just ate.

When he was feeling a little better, when it didn't seem as though bile was going to claw a hole though his abdomen, and the coffee singing in his veins was making him feel reasonably human again, he answered her long-standing question, bringing the conversation squarely back to the subject that had hovered over them like dark, moiling clouds.

"I bought a car."

She raised her eyebrows and put her coffee cup back down into its saucer.

"A *car*? Where did you get the money to buy a car? And I didn't see a car parked outside!"

"It's an old beat-up Ford Falcon and it's parked down the road, off on a side street." He dipped a slice of toast into the last of the yolk of his eggs-over-easy and took a bite. "Didn't want to be parked right in front of you, Marsha. We are linked together, and I wouldn't be surprised if you've already heard from the FBI on our relationship."

"Yes, indeed. But there's no *strong* tie between us. I just liberated you from that holding tank in Virginia. You were being held there illegally. I did nothing wrong. When they said that you were a murderer, that you'd killed a CIA agent and were being sought as the possible murderer of Captain Eric MacKenzie, I acted properly horrified and promptly promised them, Scout's honor, to turn you in even you even bothered to send me a postcard!"

"Wise move." He put a piece of bacon into his mouth, savoring its smoky, salty taste as he chewed thoughtfully. "They're really bringing out the Guard on this one, Marsha. The government wants me very badly. Apparently I can do a lot more damage to them than I ever thought possible." He shook his head sadly and stared morosely down at his meal. "What a fool I was. What a patsy, for so, *so* long."

"Hey, you haven't got time for any of that silly self-recrimination, buster. You promised to tell me about how you got that car. Where'd you get the money?"

Scarborough looked up, the memory of the past week sieving through his mind. "Credit cards. Cash advances. Machines. Oh, they've got a lot of ATM and MOST machines in Las Vegas, you bet. I charged up my limit before they got to my credit companies. Soon as we got to the quarry, I knew they'd be after me, and after me soon. It's just a wonder they didn't get me while the Arizona police were questioning Camden and me."

"About what? The *quarry*? What quarry?" Marsha shook her head. "You're going to have to start from where I dropped you off at National Airport, Ev. . . but tell me first what was at the quarry."

Scarborough pushed his plate away and grimly leaned toward her, his eyebrows lowering, a deep frown pulling the soft spiderweb of wrinkles around his eyes downward. He was tired, so tired. But he owed Marsha as much an explanation as he could give. Not just in return for this safe harbor. Because she was who she was. "That's where my daughter was supposed to be, Marsha. That's where Diane was supposed to meet her flying saucer, remember?" He looked away, his voice coarsening and dropping to a whisper. "They've got her, Marsha. They've got Diane."

"What? The flying saucer has got Diane?"

"No, no." Scarborough shook his head in exasperation. "The CIA. Or an outlaw section, a part of a conspiracy that's apparently been going on since the days of Project Grudge in the forties."

"The Editors—the Publishers? The groups that I heard Colonel Dolan and Brian Richards talking about at the Pentagon?"

"Yes. They're the ones who have Tim Reilly. And now. . ." His hand clenched around his coffee mug, for a second. Then he went back to the bedroom and pulled something out of his soiled jeans. It was a red silk scarf. "Italian," he explained. "I bought it for her as a gift when I was in Rome a few years ago. Diane loved it. Wore it whenever she could. I found it on a desert cactus back at the quarry. Tied there. It was a message from her." He sighed. "They've got Tim Reilly, and now they've got Diane." He looked at her, unsuccessfully fighting the emotion brimming up to his eyes. "I've got to get her back,

Marsha. And I've got to pay those fucking *bastards* back for what they've done to her... And what they've been doing to me for over twenty years!"

The red scarf, hanging limply from the gnarled cactus branch beneath the desert stars...

The scream of the CIA killer as he fell down the face of Hoover Dam, smearing blood on the concrete as he bump-bump-bumped his way downwards...

The crack of the gun in his hand as he and the muscular, pockmarked spook fought on the roadway above... The bullet that should have gone wild, and yet somehow caught the man in the chest...

Jake Camden's grin as they shook hands, creating an unholy alliance.....

These things, these sounds, these images, had haunted Dr. Everett Scarborough across the length and breadth of the United States as he had driven to Marsha Manning's house in Cleveland, Ohio.

And now he relived them as he told her the story.

"When I got back to the University of Kansas, I called Diane's friend's house where she was supposed to be staying. The girl said she'd gone. I knew exactly where she'd gone, too... Hoover Dam."

"Yes... where she was supposed to meet up with those aliens she and Tim supposedly contacted," said Manning. "But Ev—you explicitly *ordered* her not to go without *you*!"

"Diane can be an impulsive, stubborn woman. I hadn't shown up on time, so it would seem she barged in on Camden and demanded that *he* take her there."

"Camden! A sleazy schlock journalist! A pretty poor replacement!"

Scarborough shrugged. "Believe it or not, Camden's got his points."

"I never thought I'd hear that from Dr. Everett Scarborough, scourge of the sensationalist press!"

"What can I tell you? He saved my life. I guess that scores some points in my book."

When Scarborough reached Las Vegas, he'd rented a car and headed out to Hoover Dam.

"What I had no way of knowing was that Diane had another 'vision.' Turns out her 'aliens' hadn't meant Hoover Dam, exactly. What they meant was that they'd meet her in a less conspicuous abandoned rock quarry a few miles away. Apparently, she experienced this realization in the middle of fitting Camden with a new suit. Diane always hated Hawaiian shirts . . ."

"How feminine of her. She didn't want to be embarrassed in front of extraterrestrials by a poorly dressed companion. I can sympathize entirely."

Scarborough grunted. "Whatever. I'm not complaining. Diane was in a hurry, so she just stuck a black suit on Camden and then hurried to find the quarry. Meanwhile, I headed out to the dam to find her. Naturally, there's no sign of her, so I just wait there, admiring the view." He took a sip of coffee, remembering the faint twinges of vertigo, maybe even of latent agoraphobia, he'd felt standing there on that huge pile of concrete, a lake to one side, and a *long* drop to the other, the faint thrumming of the generators deep below vibrating his feet. "A car comes up, stops. A youngish guy with a great physique and a face like the moon steps out and puts a gun on me. Said he'd been *protecting* me all this time—now he gets to kill me. Seems like he's really enjoying his work."

"An Editor . . ."

"Right. Blue-pencilling a very large section of 'top-secret manuscript.' Let me tell you, Marsha, it didn't look very good for me at that moment."

"Perils-of-Pauline time, huh? But you're here . . ."

"Yes. Another car pulls up. A man gets out, dressed in a black suit. The car's a big sedan, black as well. The CIA guy—well, I guess we'll never know why; anyway, I *hope* not—absolutely freaks out. He starts babbling about 'the Men in Black.' This gives me just enough time to have a go at the arm holding his gun. Big fight. Turns out the guy in the black suit is Camden!"

"Diane must have sent him back to see if you'd gone to the dam for the meeting."

"Exactly. Well, I managed to get ahold of his gun and wound the guy, and I hoped to get some information out of him, but he went over the side of the dam and he's clinging there by his fingertips. I try and get ahold of him and it looks like I might be able to haul him over the edge. I call for Camden to help out. Guy gets another look at Camden—who, with typical savoir faire brings up the MIB business—and freaks again, losing his grip. Went down like a bag of rocks."

"MIB?"

"Short for 'Men in Black.'"

Marsha shook her head. "Well, your life is saved now, so maybe you can use the break for a sidebar, subtitled 'Men in Black.'"

"Saucer mythology, Marsha."

He took another sip of coffee, asked her to warm it up for him, then launched into a capsule version of his lecture on Men in Black, starting with their appearance in front of Albert K. Bender, director of the International Flying Saucer Bureau. Bender alleged that three men dressed in black suits, white shirts, and black hats, who claimed to be with a government agency, gave him some top secret information about the origin of UFOs, but warned him that they were only satisfying his curiosity to shut him up, and any leakage of the information would bring about swift retribution.

"Actually, Bender turned out to be a paranoid with a strong fascination with horror fiction, science fiction, and the occult," Scarborough explained. "Gray Barker, a former associate, wrote a couple of books on the subject which whipped up the fringe element of the UFO people to a froth in the late fifties and early sixties. Ever since, there had been *lots* of stories about the Men in Black. They usually travel in threes; they wear either well-pressed black clothing or truly rumpled. They often have slanted eyes, or are foreign looking, even though they claim to be with the CIA or the government in some capacity. Many times, they speak like characters out of a grade-B forties or fifties crime movie. They have odd areas of ignorance sometimes. What they mostly do, it seems, is ask for flying saucer information. That, and *threaten*. Oh yes, they do a great deal of threatening. And, they tend to travel around in early model black Cadillacs!"

"About an eighty-five on the Bizarre Rock Parade. Nice tune and I can dance to it!" said Marsha.

"Well, for years I thought it was absolute mythology. This Bender guy somehow linked up the old story of the devil or Traveller in Black, bearing tidings of trouble to a town, with the UFO phenomena. And it stuck! But lately . . . Christ, I don't know. I don't know anything. I mean, you should have seen that CIA guy. Really upset at the very thought of Men in Black getting ahold of him."

"Yes, yes. Noted. But go ahead with your story. Presumably, you went back for Diane to the quarry."

"Yes. Camden took me there."

"And Diane was gone."

"That's right. There were signs of a scuffle."

"What about signs of a saucer landing?"

Scarborough smiled grimly, shook his head. "Tire tracks, though. No question in my mind who took her. And Marsha, I swore then and there I'd get her back. I'm going to need your help, though."

She reached across the table and held his hand. "I'll do whatever I can."

"You believe me, then."

Her eyes lit up with a liquid brown fire. "Of course I believe you, Everett! I know you had nothing to do with Captain MacKenzie's death! I was with you when he was killed. And I'm the one who overheard the whole sordid conversation about the Editors and the Publishers."

"I just need a base, a touchstone—nothing that will jeopardize your safety or your freedom or your career. That's all I ask, Marsha."

"Okay, okay, Doctor. I *said* you've got it, for goodness sake! Have you gotten *denser* in the past week?"

"No, actually, things have cleared up a bit in my head, I think."

"Good. So go on. What about Camden? You said you're working together now. What's the story there?"

"Well, the next day we heard the news reports, and we figured out what had happened. The Editors, or whoever the hell they are, decided to pin the rap on me."

"I'd half expected something like it," Scarborough explained. "That's why I did the money-machine sweep the evening before. I figured they'd be looking for me, maybe cut off my credit, what have you. But I *didn't* think they'd pin the deaths of that agent and of MacKenzie on me!"

He'd pulled Camden out of a snoring sleep, poured a little

cold water on his head. The announcement seemed to have galvanized Scarborough, he was at a loss; now, the situation was clear: he knew he was a fugitive not only from the CIA 'Editors and Publishers' but a fugitive from the law. Fortunately, he'd not called into the Arizona police that morning like they'd wanted him to. The law had no idea where they were. When he got Jake Camden awake (at least the guy was sober and not hung over), they had a fast powwow. There had been no mention of Jake Camden in the reports. Presumably, nobody was after the *Intruder* reporter except his creditors, which left him a free agent.

"I harbor no illusion about Jake Camden. I put it right out on the table: 'You help me here, my friend, you've got yourself the story of the century! No more *Intruder* sleaze.' The road back to legitimate reporting was open to him. To fame, riches, the Pulitzer Prize. And that story was his, *if* he agreed to help me. Naturally, he agreed. There was no skin off his nose! Since he was a free agent, I wanted him to dig for information about the Editors and Publishers, and this strange program they've been perpetrating. I wanted to be able to call on him as an outside operator if I needed help. He readily agreed."

Camden was actually the one who'd bought the car—an old Ford Falcon, a thousand bucks including tax and tags. Then they'd split up, Camden catching a plane back to Florida, Scarborough hitting the highway, headed for Ohio—and then New York City.

"First thing I need," Scarborough said, "is for you to keep in touch with Camden. I gave him your number here. Any messages he has for me, he'll leave with you. Any information, papers, reports, what-have-you—he'll send here. In turn, the flow will go to him through the same channel. Is that okay with you?"

Marsha shrugged. "No problem." She was silent a moment, musing. Then she looked him square in the eye and said, "Wouldn't I be more help to you if I came along?"

"No!" He said the word with emphasis and finality.

"But isn't there someone else who can serve as your base of operations . . . ?" she said stubbornly, her brown eyes pinning him. "Look, I've got some leave coming up, and I was going to take it anyway . . ."

"Cancel the leave. I don't want you going anywhere, you hear me—not if you mean to help me. Marsha, please . . . I appreciate the thought, but I need you to stay here. For the time being. Can't you understand?"

She bit her lip and looked down at her fingernails. Scarborough noticed that she'd bitten them to the quick this past week. They hadn't been that way when he'd first met her. He hoped she wasn't biting them over him—no, of course not . . . That was absurd.

He didn't find Marsha attractive at all, never had, from the first moment he'd met her. He never much liked dark Mediterranean eyes, and her hair was much too brunette and curly for his taste. Everett Scarborough tended to go more for lighter, more northern European women . . . She was a wonderful woman, bright and alive. Spunky even. Warm and caring and all that rot. But she wasn't the kind of woman that Scarborough would ever want to have sex with, get involved with. No, indeed. They were just friends.

"All right," she said, a tang in her voice. "I understand. Charles Bronson has got to go out and be Mr. Vigilante. Solve the crisis on his lonesome. Walk tall and all that."

"Marsha, I'm not on my lonesome," Scarborough said, reaching out and taking hold of her hand meaningfully. "Not with you back here, helping me."

"Oh, right," she said, pulling away from him, getting up and walking on her brightly waxed linoleum floor. She spun around, one hand on her hip, one forefinger cocked like a mock-gun, pointing at him. "He gets his stomach fed and suddenly the old, self-efficient Scarborough rears his blocky head!"

"Oh, thanks for the sympathy!" He felt a little stung.

"You've got sympathy from me, you've got empathy, you've got everything you need, Everett! Why don't you take *more* when it's offered! You're in way over your head. I've been in active service with the military for over ten years. I *know* things; I can *do* things."

Scarborough sighed and sipped his coffee calmly. "I'm sorry, Marsha. You know, I really *would* like to have company."

"Company! Haven't you been listening to a word I've said! I'm one competent lady, pal. You know what they used to call me in ROTC back in college? Lieutenant Bitch. Behind my back, of course, 'cause they knew I'd beat their butts if I heard them. I got my black belt in karate in high school, I'm working on a blue in aikido now—and I'm a top-notch shot. You show me a computer network and I can wiggle in . . . You're just like my superior officers! You just won't give me a chance!''

"I don't doubt any of that, Marsha. But I don't care if you were your class valedictorian—''

"Just missed it by a grade-point!'' she said defiantly.

"You could be Supergirl, and I wouldn't care, Marsha. You seem to forget that you've already risked plenty when you sprung me from that military brig in Virginia.'' He shook his head and wearily put his head in his hands. "Don't think I haven't thought about this. I thought about it all the way across nine states. This is my battle, Marsha. I'm asking too much of you as it is. I can't allow you to risk your *life*!''

"Isn't that *my* decision, Everett! Isn't it *my* life? Look, do you think I'm particularly happy here in the Air Force! Let me tell you, I'm not! You think I have a whole lot of respect for Washington, D.C., for the Pentagon. No way! And my career! I'm just stuck in a goddamned rut! Helping you out has been the first meaningful thing I've done in, I don't know *how* long!''

She turned away from him. Was that a cough he heard, or the stifling of a sob? He got up from his chair, went to her, reached out, and touched her gently on her back.

"And you'll be helping me out this way, believe me, Marsha.'' He said softly. "Look, I promise, if it gets to be too much for me . . . and I need help. I'll call. Okay?''

She nodded. "I'm sorry. I guess I'm being overemotional about this, Ev.'' She turned and looked at him, frowning. "I just can't get Richards and Dolan out of my mind. Two high-echelon officials Ev—talking like *thugs*! Working for some intragovernment conspiracy to commit God-knows-what atrocities! When I took my wings, I promised to protect my country—and when I work against those bastards, that's just what I'm doing.''

"Well-said, Marsha, and I do understand your disillusionment and anger. I'm brimming with those two emotions myself, believe me.'' He chuckled ruefully. "I'm going to put things right, Marsha. I'm going to find out the truth, and get that truth out to the people. And finally, I'm going to find my daughter."

She nodded, smiled, and put her head on his shoulder. He could do nothing but hold her, feeling her racing heart hammering against his chest. She was so warm, and she smelled so damned good . . .

But no, he wasn't attracted to this woman.

She was a friend.

A very close, meaningful friend.

That was all, dammit!

"But what are you going to do now, Everett?" she asked. "Where are you going to go?"

He told her.

CHAPTER FIVE

The woman pulled off the earphones from the typing man's head. Blues music squealed tinnily from the tiny speakers.

"Here are your phone messages, Jake," said Betty Norton throwing a stack of torn-off, pink paper memo-slips down on the disaster-area desk. "It's after six, pal. I gotta go."

Jake Camden squinted up from his rattling IBM selectric, the Camel butt dangling from between his lips like a punctuation mark in a Tom Waits song. "Sure, babe. I think I gotta few more pages in me tonight, though. I'm gonna stay late."

Betty Norton was the assistant he shared with the Human Interest editor of the *National Intruder*. Jake Camden was the paper's UFO reporter. He was supposedly working hard now because he had several weeks of columns and stories to finish before he shunted off for further "field work," as he'd described it to his boss. Actually, he'd finished his *Intruder* work yesterday. What was in his platen now was a first draft of the story he was going to sell to the *New York Times* or the *Washington Post* or maybe a syndicate—any venue with the class and respectability that a rag like the *Intruder* lacked. It was going to be his ticket out of here, this story about Everett Scarborough. Certainly he couldn't show it to anyone yet, tough—he and Scarborough needed to dig up some more documented fact. But when they did! Whew! Wow—eee! It was gonna be huge! He'd get a book deal, a movie deal, and most important of all, a ticket out of Crackerville, Florida, humid home of Jaundiced Journalism. He'd have his dream . . .

"I've never seen you work so hard before," said Betty, shaking her head.

"Things are hot, babe. Real hot!" He turned back to the keyboard and started pounding. After a moment, he sensed that she was still there. He squeaked his swivel chair around, took the Camel out of his mouth. "What are you gawking at, sweetheart!"

"You haven't made any lewd suggestions or asked about my sister all week, Jake. What's going on? The doctor putting you on 'human' instead of hormone shots?"

She was a bottle-blonde, with sufficient figure and a cute enough nose to merit Jake's scattershot approach to nabbing women. He'd nailed her sister, Ann, from Tampa on a visit last year, and that was close enough. But he kept trying anyway. He figured she'd be hurt if he didn't. However, ever since he'd come back from Las Vegas, he hadn't given *any* women the usual Jakey business.

There were more important fish to fry.

Jake smiled at the jibe. "I know you miss the attention, kiddo, but what can I say? The saucer men aimed a work-ray at me. I'm dancing at the end of extraterrestrial strings! I gots to lets mah fingahs dance on duh keyboard!"

"Hey, I'm not complaining, Jake. Don't get me wrong. Goodnight." She started to leave.

Jake rifled through the messages, suddenly remembering who might have called.

A return call from MUFON. Something from the Air Force–public relations branch. The usual cranks with their stories of flying saucers shorting out their television reception.

But nothing from Everett Scarborough, nothing from anyone associated with Scarborough.

Jake got to his feet, working out a crick in his neck. Half of his rumpled Dacron blue mangos and green-parrots Hawaiian shirt dangled out from his Sears Sta-Prest pants, half was sort of tucked in. "Betts! Just a second! Yo!"

She turned away, warily keeping her distance, looking uncomfortable against the panorama of empty desks, chattering AP and UPI machines, and scattered papers that composed the boiler room of The Paper That Cannot be Avoided. Jake's office was little more than a cubbyhole against a wall, formed of nose-high painted sheet-metal and cheap tinted glass. All hail Kozlowski, generous all-Father, and cheap ruler of the *Intruder*. Actually, Camden was grateful for what little privacy he got—especially in the present situation when he *needed* it.

"What?"

"Nothing from Ed Chaney?"

"You keep on asking that, Jake. If you got a call from Ed Chaney I'd have put him through, per your instructions. Now can I *go*, my liege. I *do* have a boyfriend, and he is aware of the sleazy reputation of a certain snake I work with!"

Jake smiled, pleased to have the admiration of his male peers. "And Kozlowski! What about Kozlowski? He say anything about my articles?"

"Nada. Adios, Jake."

"Thanks, sweetheart. Say hello to Billy-boy for me. Tell him I'm sorry I threw up on his golf shoes at the barbecue!"

Betty's Nikes were already squeaking down the Formica Five Hundred, her rear wiggling prettily. Jake could not help but watch the view for a moment, sucking his teeth appreciatively.

Jake Camden thought of himself as Spencer Tracy reborn in a scaled-down version of Tom Selleck, but the general, movie-star associations pinned to his appearance and manner suggested that he was the bastard son of an unholy dalliance between Mickey Rourke and James Woods. What the hay, anyway! All those guys got laid regularly, and so did Jake, with a comely variety of the wimmens . . . That was what mattered, wasn't it?

Camden had a sallow, unhealthy face, saved by a narrow chin and high cheekbones and eyes that could shoot a bird off a branch at fifty yards—when he was sober. When he wasn't though, the eyes got all half-lidded and bedroomy, and the chicks liked that, too. He had mousey brown hair starting to recede in a widow's peak, and a mesomorph body beginning to bloat up a bit in the midsection from his drinking. Camden didn't worry though—he could take it off in a few weeks of racquetball. Right now, Camden looked like hell, stubble on his face, rumpled, his hair disheveled. But he always looked like that after a hard-day's work, and he was proud of the evidence that he'd actually spent more than the obligatory couple of hours behind the typewriter.

He went back to this desk, popped the cassette, and put a B.B. King tape on, fitting the earphones over his head snuggly, thumbing up the volume.

Sing to me Lucille! Sing me dah blues!

He tapped a Camel unfiltered from a crumpled pack and fired it up, downing the dregs of cold black coffee from a Styrofoam cup. Then he glanced over the last story he'd worked on and gotten stuck on yesterday. He wanted to hack out an ending real quick, so he could leave it for Kozlowski's inspection the next day.

DO UFO'S STEAL YOUR GARBAGE?

Mounting evidence all across the United States has begun to paint a harrowing picture of a new and frightening trend to emerge from the UFO scene. Increasing evidence of aliens invading the privacy of American citizens again and again, from Bangor, Maine, to San Pedro, California, has recently come to the attention of this reporter, making him ask a simple but profound question:

Do you know where your trash goes?

Do extraterrestrials abscond with the cast-off minutiae of your everyday existence for their own inscrutable purposes?

Harold Budkey of Omaha, Nebraska, tells an unnerving, but increasingly prevalent, story:

"It was about three-thirty in the morning and I was sleeping when I heard the most confounded banging noise. I thought it was cats or dogs getting into my garbage cans, which I'd put out in the front yard for the trashman to pick up the next day. I put on my bathrobe and I looked out the front door, and damn if

I didn't see this little bald guy, about four feet high, rootin' through the can. And hovering overhead was a flyin' saucer!"

The question is, What do men from other planets want with coffee grounds, TV-dinner packages, and empty, smelly cat-food cans . . .

Yeah, thought Camden.

Just exactly what *did* creatures from other planets want with people's garbage?

He leafed through his pages of notes. Actually, he had about three reports of this particular saucer phenomenon, culled from his six years of investigations in the field. A little exaggeration never hurt business . . . though. Besides, he had to crank out some reserve stories . . . some *good* stories, rife with middle-American paranoia, to tide him over while he went out to do the stuff that Scarborough needed him to do . . . And to dig for the information his hard-hitting exposé needed.

That, and to talk to Maximillian Schroeder.

An idea suddenly struck him like lightning. Maybe there were some kind of DNA remnants of human beings that aliens dug out of the garbage! Yeah, and they used this to form clones of people in their saucers . . . No . . . no. On the Dark Side of the Moon! That's where they did it. Appropriately inspired, Camden began to scribble out his volley of thoughts on a legal-sized pad of yellow paper. This was what he was doing when Kozlowski found him.

Camden almost jumped out of his socks when the Old Man tapped him on his shoulder. He pulled off his headphones and wheeled around to stare up at the squat, round man, peering down at him through thick glasses, the ever-present foul cigar jutting from big lips like an accusing finger.

"Sir!" said Camden, his surprise forcing unusual civility instead of wisecracks. "I didn't know you were still here.

"Gotta work hard to keep this mill grinding, Camden!" Kozlowski said, scratching a particularly ugly and prominent mole on his neck. "Just took an important meeting. Bit of a problem with that business about the Reagan White House Orgies story. Looks like I'm going to have to tell Blake to tone it down some."

Camden leered. "Too bad, Koz! That was one hot story!"

"Why don't you give me an orgy story, huh? Yeah. Orgies on flying saucers! We shoulda thought about it earlier."

Kozlowski's corpulence was of the greasy variety, the sort reserved for those truly ugly souls of the business world who wore their iniquities upon their bodies like cartography. He smelled of Old Spice defeated by fungoid growths in the damp parts of his body, and of the onion-stocked grinder he must have just snacked on during that meeting with Joe "the Carnivore" Donohue, the *Intruder*'s lawyer who was rumored to moonlight in *Jaws* movies. Kozlowski spoke with an accent from the wrong side of the East River, and he was rumored to be destined for a cement raincoat-grave in that body of water if he ever ventured back north.

"Sure, boss. I'll get on it right away." Camden kept the smile pasted on his face. "Those other stories I gave you couple days ago . . . Whatcha think?"

"That's what I stopped by to talk to you about, Camden," Kozlowski chewed on his cigar a moment, letting his reporter squirm for a few moments. "Not bad. Not bad at all. I see that our little talk about your performance here rattled some sense into your head. Or maybe a little inspiration into your typewriter, huh. I especially liked the story about compact disks containing secret messages from extraterrestrials. I think we can run that, and then use it again with secret messages from Elvis Presley next month. Whatcha think!"

"All yours, boss! Thanks." Camden cleared his throat. "So, if you, er, liked the stories, Koz . . . Well maybe . . . I thought . . . uhm . . . "Geez! He was hemming and hawing! Camden usually never did that—Koz really had him rattled. "Maybe some kind of bonus might be in order."

The eyes immediately become slits. "What . . . you want more *money*?"

No, you idiot! I want a box of your cigars!

"Sure! I find myself a little overextended. An extra bit of cash would sure help me out."

Kozlowski was quiet for a moment, the wet, slobbering cigar working around in his lips and teeth like a dog chewing on an eel. Christ, though, this was a change, thought Camden. He wasn't spitting out an immediate "No, you asshole!" like he usually did. What kind of evil plot was cooking up there in that foul kettle of worms Kozlowski used for a brain?

The beady eyes wandered over the piled-high desk, to the old IBM rattling by its side.

"Watcha writing there now, Camden?" he said, craning over to read.

Camden's first reaction was to hide his work from his boss, but he cancelled the impulse immediately. That would excite Kozlowski's interest. He had to be cool. Besides, the Old Man couldn't read print unless it was write up against his nose. "Oh, just finishing up this piece right here." He shoved the first part of the Trash Can piece at the fat man, who rattled it close to his face to check out the headline.

"Hah! Great, Camden. Good stuff! Yeah, that's personal, all right, just like I want. That'll get John Doe checking his backyard for the critters from Jupiter." Then his thick eyebrows squirmed together like mating caterpillars. "But what's the reason, Camden? Why *would* ET's wanna steal garbage?"

"DNA, Koz!" said Camden, rejoicing in the serendipity of his recent musings on the very same subject.

"Dee En Ay? What the H is that?" Koz chuckled.

Camden considered a quick genetic lecture, but then decided against it, bringing into play a more recognizable word. "Clones, Mr. K. The stuff they make clones out of. Maybe the aliens are making clones of average American citizens upon in space—or on the Dark Side of the Moon."

"Yeah!" The cigar waggled enthusiastically. "Yeah, I like it a lot." He tapped the paper with a thick forefinger. "Hey! You think that's where Elvis has been hidin' . . . the Dark Side of the Moon?"

"Could be, Koz. Could be. But then, maybe he's on one of the moons of Jupiter too . . . maybe Callisto. Lots of UFO activity around Callisto."

"Great. A whole series. I want you to have a meeting with our Elvis editor, as soon as he gets back from doing that piece on the Graceland sewage system."

"Sure, Koz, no problem." He didn't bring up his impending trip. Koz would probably forget all this, anyway. But there was one subject that Camden didn't *want* his boss to forget.

"Really, though. What do you think? This stuff deserves a little extra green to keep the author happy and flowering!"

Koz leaned against the desk precariously, and dug a finger into a nostril thoughtfully. He examined what he'd dug out, pulled out a handkerchief, and deposited his prize amongst the day's collection. "Money, huh, Camden? You're really somethin', you know? A week an' a half ago, you were just happy to keep your job after I almost tossed your tail into the parking lot. Now, you want some gravy." He nodded to the typewriter and tapped the beginning of the Trash Can piece on the desk, slightly smearing it with snot. "This stuff—it's okay. I like it. But you ain't gonna win no Putz Seltzer prize for it!" The Putz Seltzer prize was Kozlowski's mythical award for best supermarket tabloid journalism of the year.

"Yeah, Koz, but come on! Week after week, year after year, I gotta churn this stuff out! I gotta thin it a *little* to stretch it out."

"I ain't told nobody, and maybe I shouldn't be tellin' you, Camden. But the *Enquirer* and the *Star* pulled another *million* ahead of us last year. We ain't lost no readers . . . There's just more that we can get. That's why I'm on you for some hotter pieces. We want those extra readers." Kozlowski grew quiet. Camden thought for a moment the man was contemplating some deep philosophical announcement, but then a sudden eructation, redolent of oregano, pepperoni, and oil, issued from his jowly face. "Tell you what, Camden. Don't ask me why I'm doing this, after that business with you and my daughter. Your balls should be on my wall, I guess, but I'm a forgiving man. I'll make a deal. You come up with somethin' earth-shakin' . . . some great story that will make those million readers pass over the other rags and stick the *Intruder* in with their milk and eggs and bread, and I'll give you that bonus, to the tune of ten thousand smackeroos."

Camden blinked. "What's the going exchange rate on smackeroos, boss?"

"Last I looked, it was one American dollar per . . . I see you're surprised. Didn't think I was that generous, huh? Well, it's gotta be good, Camden, so I probably won't have to give it. And the way I judge it's good is, I get that million circulation increase, the issue of the *Intruder* it's in. That tantalize you?"

"Sure. But that doesn't do me any good now."

"I see a possible, Camden, it's an instant *two* thousand, cash, no deposit, no return. How's that grab you?"

"Well, I guess I'll start working on it, Koz."

Kozlowski patted his employee on the arm. "I knew you'd get real motivated when you heard that, Camden. And I wouldn't be making you this offer if I didn't think I could squeeze an incredible story out of you." He stood up and began to waddle off, leaving a pall of cigar smoke behind him. "You know where I am, Camden. Dig me up some great stories, I dig you up some significant green."

Camden dropped down into his chair. He reached for the coffee cup for a drink, but it was empty. His heart beating quickly, he got back up and headed for the refreshment station, a little nook on the other side of the room.

Twelve thousand dollars, with two upfront. Snap! Whew, he sure could use that money!

The thing was, Camden knew that he had just the story. And he was sitting on it!

The refreshment station sat recessed in an alcove, a Coke machine was humming in a corner, and a Mr. Coffee machine was sitting on an imitation mahogany shelf. There was about half-an-inch's worth of deep black liquid left in the kettle, and when Camden poured it, it slopped out thick and *burned*. He barely noticed. He was too busy thinking.

Geez, what would be the harm? He didn't have to give the *Intruder* the full story. He could make it a *series* of articles, yeah . . . The first few, vague and unspecific. The first wouldn't hit the stand for a few weeks . . . By the time the series was over, he could have sold the *real* piece to a respectable paper, along with book rights, and he'd be outta this dump! And Scarborough would have found Diane by then, surely, and this would be all over . . . What was the harm?

He sipped the bitter coffee and cringed at the dreadful taste. He diluted it with some water. Still bad, but drinkable at least. The caffeine quickened his heartbeat as he thought; and then came the memory of Scarborough's hands on his collar, Scarborough's face in his face: "I better not see word one of this in the *Intruder*, Camden. If I do, all deals are off. No exclusive. No collaboration on a book. And remember, I'm wanted on suspicion of murder now. You print a hint of this in that shitty paper of yours, my friend, and maybe I'll come visit you sometime and make good on that suspicion!''

Unlike Dr. Everett Scarborough to make threats of violence like that, but then Camden had never seen a man change so quickly from a civilized sort to a driven creature. Of course, having a best friend killed, getting your daughter kidnapped by an outlaw branch of the CIA, and learning in the bargain that you've been a patsy for over twenty years wouldn't put anybody in a good mood.

Naw, thought Camden. Maybe that's not such a good idea.

Anyway, he liked Scarborough, he really did. A nice daughter too, a real fox. What's more, he'd actually gotten along very well with Diane Scarborough's boyfriend, Timothy Reilly. A good guy. What with Tim and Diane imprisoned somewhere, undergoing God knew what, it probably wouldn't be a real good idea to rile the bad boys of the Company Annex up.

Camden sighed. Oh well, he was getting out of town soon enough—he had a seven-thirty USAir evening flight out of Orlando tomorrow for New York. Maybe he wouldn't need that quick cash after all. This business with Max Schroeder might pay off.

On the table by the Mr. Coffee was a Maxwell House coffee container bannered with the sign: Coffee Contributions. 25 cents suggested. You Too, Jake.

Purely out of habit, Jake Camden looked behind him to make sure nobody was about, then poured out the collected quarters in the can into his palm and pocketed them.

Enough work for today, anyway. He needed a drink, goddammit!

Too much of this virtue nonsense was making a boring guy out of him. A guy had to have *some* fun!

He went back and stuffed the Scarborough article into a briefcase. Not that he intended to work on it at home by any means—he just didn't want it lying around here at work, available for the many snooping eyes of his colleagues.

Just as Jake Camden hit the door and pounded down the stairs, eager to hit the Palm Branch Lounge, the phone on his secretary's desk began to jangle.

However, Jake did not hear it ringing—his thoughts were

turned toward a few glasses of joy, a couple games of pool, and maybe some pretty tourist on her way down to Disney World.

CHAPTER SIX

Goddamn it, *answer the phone*!

Everett Scarborough let it ring for the eleventh time and then hung up. Coins clattered down into the public phone box.

Well, he should have called Camden earlier.

Wait a minute. Didn't he also give him a phone number? There it was, on the other side of the hotel stationry. Scarborough put the coins into the slot again, got the long distance operator, and placed the call. He didn't dare use his Sprint FON card, for the same reason he couldn't use his credit cards. He was dealing with high-tech people here, and he couldn't rely on things that he used to take for granted. They could trace him too easy.

Everett Scarborough, low-tech refugee from justice. If he survived this ordeal, Hollywood could use this story for a series. Too bad David Jannsen was dead—but then, Don Johnson might do.

The operator came on and helped him with the call.

The connection was made on the third ring.

"*Que pasa!*" came Camden's voice, hoarse from too much booze and too many cigarettes. "I can't come to the phone now, I'm setting up my machine-gun nest to deal with burglars. If you saw a green man from Mars, call my secretary at my office. If you have money for me, leave your name, address, and phone number. If you *want* money from me, come on over. I wanna test out this machine gun. Oh yeah, in just a very short time I'm going to say a filthy word and the phone company's gonna beep me out. You got a message, let it fly then."

BEEP!

Scarborough hung up.

If he got the chance, he'd try later.

He was outside a 7-Eleven store in a town called Imperial, close to Pittsburgh.

Everett Scarborough had been traveling since about two in the afternoon, still keeping to back roads, still toeing the speed limit, still keeping the lowest road profile possible. Now that he knew exactly where he was heading and what he was doing, his pace was frustrating. Before, he hardly noticed. He was too busy thinking, weighing the situation, and above all trying to get his emotions in control.

He went inside the store and bought a paper and a Snicker's Bar. He'd have some dinner later—right now, his low-sugar level needed a boost, and so did his mood. At home, in this kind of funk, it would have been scotch time. Now, he couldn't afford the luxury of drinking.

"Hey—haven't I seen you before?" said the clerk, a skinny guy with limp, long hair and a large Adam's apple.

Scarborough shrugged, hiding the sudden lunge of his heart. Before, he had luxuriated in recognition. Now, he cringed at the very thought.

"I don't know," he mumbled, "Have you?"

The young man squinted at Scarborough for a long time. "Naw, you just look like him, I guess."

"Oh? Who do I look like?"

"This state official I seen on TV. Can't remember his name. Thanks." Change was handed over, and the guy went back to studying the black-and-white screen on a portable TV, its aerial augmented by Reynolds Wrap.

Scarborough grunted and left.

He went back to the group of three phones to the right of the entrance, by the large icebox. He was going to try another call—a call to a man he'd counted out at first, but now, upon reconsideration, who just might be willing to help him.

He pulled his pocket address-book out of his back pocket, consulted the section marked 'M', and started to dial a 703 area-code number. He remembered Ed's voice just last year as he pushed the business card into his hand. "Ev, thanks. I owe you one. You ever get into trouble, here's my private consulting number. Phone in the basement, separate line—registered pure and untapped."

Edward Myers.

Ed worked out of Langley, Virginia, but he wasn't in his office very often. He tended to travel a lot, training division-heads and generally keeping the trench-coated men up on the latest in hardware, software, and wetware pertinent to their calling. In short, he was a career spook. CIA. They'd met at a think-tank sponsored by the government, brainstorming spy-satellite operations. Scarborough was an old hand at this, way back since his days at MIT.

As soon as Scarborough started making significant money as the country's leading UFO skeptic (beating out even Phillip Klass for that title, mostly because he was much more photogenic and charming), he did very little consulting, only enough to keep him abreast in significant aeronautic developments.

He and Myers had hit it off at the think-tank. Myers had come out a few times for one of Scarborough's infamous weekend-long poker games. He'd met Mac MacKenzie there and even Mac thought he was a neat guy, for a spook.

Their relationship would never have gone any further than poker games and dinner parties (Cynthia Myers was forever trying to fix Scarborough up with some woman), except for David Myers, Ed's son.

David had, early in life, discovered the entrepreneurial tradition of free enterprise that had made America an economic giant. Unfortunately, David sold drugs. Busted at age sixteen, addicted to coke, various upper and downers, and possibly to alcohol, David had only been kept away from serious discipline by his father's government connections. Myers got him treatment, but another ingredient was necessary—attention. The CIA agent was gone so much for his job that he never gave David the attention he needed. He tried to make up for that, but he also encouraged David to reach out to other older people as role models. One of David's sober enthusiasms was science, and one of his favorite authors was Everett Scarborough. Every Saturday afternoon for close to a year, Scarborough had either visited David or had David over, and personally guided him through a tutorial of friendship, reading, and private discussion.

David had proved to have quite a facility for computer

science, and that last thing Scarborough had done was to teach him the basics of BASIC, FORTRAN, and C. Now, David was entering the University of Virginia, drug-free, addicted only to computer programming, girls, and rock and roll. Myers had been very grateful to Everett, certainly; but actually, Scarborough did it because he liked David, saw a lot of himself in the boy. Besides, it had all been fun, and he'd gotten a chance to try out a lot of potential lecture-routines aimed at the high school and college age. A healthy percentage of his yearly income came these days from the lecture circuit. Scarborough had never intended to call in the favor.

Until now.

"This is Potomac Consultants, Inc." came Ed Myers's answering machine tape. "Please leave your name and number at the sound of the tone."

Much different and certainly more professional than Jake Camden's message.

Should he leave a message? Yes, Scarborough decided. He had to take that chance.

"Ed. It's me. I'm sure you know, I'm in deep trouble. Maybe you also know I've been framed. I need your help. ASAP. You can reach me through my agent, William Franklin."

Scarborough rattled off the appropriate 212 number and then hung up.

When he started walking back to his car, he immediately saw the Pennsylvania State Trooper car, and the policemen looking over his car. His heart leapt up to his throat, but he contained the urge to run. Without breaking stride, he continued toward the car. If the cop asked for a driver license, maybe he was screwed, maybe he wasn't. But if he ran away, the jig was up.

"Can I help you, officer?"

"This your car?" said the man, a pudgy sort, bored-looking, doubtlessly counting his months to retirement. He barely glanced at Scarborough, which was an excellent sign.

"Yes."

The man in grey lifted a scuffed shoe and tapped the bumper of the Falcon, below the headlight. "Got yourself a busted turn-signal lens, Mister. If I wasn't off-duty, I'd write you up. Take my advice. Get it fixed."

"Oh—ah . . . thanks officer. I hadn't noticed."

The officer grunted, hitched up his gun holster, and waddled into the 7-Eleven store.

Scarborough breathed a sigh of relief, bending over and examining the half-broken-off red signal-cover.

First body-part shop he found, he'd either get a new lens or something to disguise the breakage. He didn't want to get caught by the authorities for something so petty!

He got into the car and, quickly as he dared, got the hell out of there.

CHAPTER SEVEN

There was a new bartender at the Palm Branch Lounge. Female. Pretty, and not a bad body. A little mileage on her, but Jake Camden liked her hairstyle, long on the sides with her bangs moussed to either side of her part in the middle.

"So I says to Steve Spielberg, 'You get yourself a decent script, we'll talk business!' " Camden drained the rest of his Moosehead beer, as much for dramatic reasons as for the drink, and then looked deep into Cover-Girl-adorned eyes, mustering every bit of his sincerity. " 'Steve, I hate to say it, but *Close Encounters* and *ET* were cute, but they were about as close to the truth as Disneyland is to Harlem. You decide to make another film about aliens, it's gotta be hard-hitting and realistic.' I gave him *my* ideas, and he's getting his writers working on them. They're what they call 'in development,' in Hollywood parley. They're paying me some nice option money, but who knows what will happen. People don't want the truth, they want pablum. I don't deal with pablum, I deal with *reality*. Speaking of which, Annette, how about a nice brimming shot of Johnny Walker and another bottle of this Canadian stuff?"

The Palm Branch Lounge, a seedy bar with an air of faded

gentility, hung over a canal, shielded halfheartedly from the Florida sun by a brace of anemic palm trees. The big excitement for Palm Loungers was whenever Gertie, a nine-foot-long alligator, showed up below the deck; the regulars would feed her barbecued potato chips and marshmallows. The marshmallows were particularly amusing, since they tended to stick to the long teeth. The Palm attracted a lot of well-to-do alkys and bored wives of out-of-town businessmen, which made it a primo shot for pickups and cons for a slick article like Jake Camden.

"That'll be three-fifty, Mr. Camden," said Annette.

"Jake, kiddo! Call me Jake, please!" Camden pulled his wallet out, stared at it mordantly. "Gee, Annette. I'm tapped out. Forgot to stop at the Money Mover! Can you put it on my tab? I'm a regular here—but wait a minute, not now . . . You wanted to hear about how I got into flying saucers, didn't you." He bent over conspiratorially, as though he was about the tell her the Secret Knowledge of Everything. "Have *you* ever been picked up by aliens, Annette?"

Annette blinked. "Hell, no!"

"Are you very, very sure? You know, they can erase your memory of the event. But there are signs . . . Indications. You want me to give you a few, just to make sure you're right, just to make sure you haven't been secretly?"

"No way, hon. My necessary equipment went to Mastectomy City!"

"Just joshing with you, Annette. Seriously, though, you want to hear my story?"

She eyed him suspiciously. She was a lot cuter when she was smiling; she lost her dimples when she frowned. "How come, Jake, I get the feeling these phrases have tumbled from your lips before?"

Jake studied her a moment carefully. Wait a sec. Yes, the exterior was hardcase enough. This lady had been through enough men and pain to fill a year's worth of *True Confessions*. But she couldn't hide the spark of interest in those still young hazel eyes of hers.

"It's not just my job, Anny . . . The serious journalistic pursuit of Unidentified Flying Objects is my life. You can be certain that I too, was doubtful at first. The truth is that over 95 percent of sightings of flying saucers can be explained as natural phenomenon, delusions, or hoaxes. But what about the three or four percent that *can't* be explained? That's the question I always asked myself, when I first started doing a few stories for the *Intruder* on the subject."

"What, you didn't see one yourself?" The bartender unwrapped some Wrigley's Spearmint and stuck it into her mouth.

"Naw. Maybe one day. But you know, it doesn't make that much difference. All this stuff is so much a *part* of me . . . When I hear, first-person, a genuinely sincere encounter of a sighting . . . well, babes, it's amazing. I'm just swept off my feet. *I'm there*!" Excited, he got off the bar stool, executing dramatic gestures with his hands. "When they tell me they see saucers hovering above power lines in the Midwest, suckin' up that energy to stoke their energies, I'm there. When John Doe tells me he's seen short, funny-lookin' men peering in through his bedroom windows and later a round shape pulsing with red and green lambency—shit, I'm *there*! When Jane Doe is cryin', telling me how the aliens are showing her a little child, genetically half-terrestrial, half-otherworldly, yanked from her womb, taken from her loving arms . . . shit, Annette. I'm *there, too*! Blubbering away, often as not!"

Her features were softening. He was getting to her.

Maybe she'd even forget to write his tab down in the tattered black book that Fritz, the owner, kept by the cash register.

He told her—and this was for real—he was genuinely fascinated with the world of UFOs. He'd plunged into the wealth of literature on the subject—in print, or out of print . . . He'd actually made a trip to the Library of Congress, Washington, D.C., early in his career, just to research there for a solid week. In ringing phrases, he described the sense of awe, the thrill of wonder, he'd felt at the possibility that other life might exist on other planets, and that intelligent beings might actually be observing humans, even secretly mixing it up with us. What Jake Camden did *not* tell Annette was that this buzz lasted a whole year and a half, before he started to realize that all this stuff was in people's heads. That, in all likelihood, it was all hokum. But he was too far in by then, making too much money, getting too much attention to back out. By that time he was already working on that cheapy film, the one with the caricature of Scarborough in it.

He'd been hooked.

"Oh shit" said Annette. She was looking down into the black ledger book. Camden had been so busy expounding upon his dramatic, exciting, crisis-ridden life he hadn't noticed her getting the book. She gave him a hard look. "You asshole. Why didn't you tell me you've been cut off!"

Jake blinked. "Cut off! What, my tab? Here, let me take a look at that. There must be a mistake!" Nonetheless, he quickly finished his draft, even as he looked down at the winged open book. No mistake, of course. He owed Fritz a good three hundred smackeroos, and in big block letters, the heiny's message was clear: NO MORE TAB FOR THIS BOZO. ALL CASH!

A couple of other regulars at the end of the bar hovering on their stools like vultures behind wavering streams of cigarette smoke chuckled. They'd been waiting for this little confrontation all along. "Hey, Jake!" one of them yelled down. "You owe me ten bucks, too."

"Twenty here!" cried the other.

"You guys will get your goddamned money!" he yelled, then sweetened up as he turned to Annette. "Tell you what. You spot me for these drinks, and I'll take you out for dinner and a show down in Orlando when I get back from New York."

She glowered at him.

"Okay, okay! What about a weekend in Palm Beach. I know this cute little UFO-nut couple down there. Lovely house . . . I'm welcome any time!"

"I can't afford these drinks! I'm gonna lose my job, you shit!"

"No, no honey. Fritzy won't know."

One of the regulars, a wizened old guy with a cackle like a crow, piped up. "Oh yes he will, Camden."

"Okay, okay, I've got my money machine down the street. I'll see if there's anything left in my account besides fumes!"

He got off the stool. There was a little money in his checking account, sure . . . enough to pay the bill at least. But he needed it, and no way was he gonna come back and pay this broad. Besides, she wasn't that good-looking anyway. Jake had been getting spoiled on juicy college babes, ever since he'd been booked into that speaking tour last year. Yeah, what do you want to fool around with a dishrag like this, he told himself. Next time you come in, you give her the money and a nice tip, along with some of Fritzy's money, too. Hey, everything will be absolutely jake then.

Of course it probably wouldn't, but by the time Jake hurried out of the bar, followed by the new bartender's hurled obscenities, it didn't make much difference. Part of the secret of Jake's ability to seem sincere was that he loved the sound of his own voice so much that while his mouth was yapping, he believed everything that came out.

Weaving a bit, feeling glad to be out of that particular situation, and with a nice little splash of free booze in his system, Jake went around to the back parking lot to get his car. The drive back to his apartment was short; no danger of some cops laying a DWI on him. (He'd gotten two of those last year, and probably would have served prison time, but Kozlowski had the local constabulary in his pocket and he got off with probation before judgment, small fine.)

They were waiting for him by his car.

Camden didn't see them at first. He was already fumbling with his keys, whistling "Eternal Flame" by the Bangles and wondering if Susanna Hoffs was into UFOs, when the three men stepped out of the shadows by his 1988 Ford Escort.

He didn't recognize any of them at first, and the pleasant insulation of the alcohol from the Palm Branch Lounge made him slow to sense danger and even slower to react. Otherwise, he might very well have turned and immediately run back into the bar, irate bartender or no.

"Jake! Jake, my man, *mi bueno, amigo. Buenos noches.*"

A short man wearing a tank top below a Banlon jacket, black chinos rolled up to show off white Air Jordan basketball shoes, separated from the other, tall, thicker forms. Jake recognized the voice immediately, and cringed.

"Oh!" he said. "Hey, Johnny. Hello! I've been trying to get ahold of you!"

"Yeah? Is that right, Camden?" The man's hair was dark black and gleamed with Brylcreme. Something splashed in the nearby canal. A passing truck roared by on Mangrove Avenue. Camden could smell the pungent odor of marijuana.

They'd been out here, waiting for him for a while, and if they had anything civil to say to him, they would have come in and bought him a drink. "I didn't get any message from my answering service, man."

Johnny Plentenos spoke with a pronounced Cuban accent, and a sort of a Desi-Arnaz-on-downers slur. Camden had heard him speak unblemished American English when it was necessary and he claimed he had a degree from some Havana university. When he used his Cuban accent, that meant he was ensconced on a duty of his drug trafficking that he either found distasteful—or enjoyed very much.

"Johnny, I'll be straight with you. Like I said, it's going to be another little while before I can get you your money. Have a little patience, okay?"

The other two men stepped out into the light. They wore black t-shirts, the short sleeves bulged out by biceps the size of pregnant melons. They had boxers' faces, broken noses, and small eyes which devoured Camden hungrily. One had long hair, the other had short, but otherwise they looked like clones from the same aboriginal vat.

"Patience! I *am* a patient man, Jake. I been patient with you for what, two months now?"

"Look, I got you some money last month and the month before. Hell, Johnny, I sold my IBM PC toward paying you off. Give me a chance, huh? I'm onto something big, and I'll have your money in two weeks!"

"Two weeks! I can't wait two weeks, Jake. You don't know? This ain't no credit operation the Colombians run? Me, I'm a nice guy, but I don't get them their money, Jake, they cut off my *cajones* and stuff them down my throat!"

He snapped his fingers and the muscular boys stepped forward toward Camden, smiling.

"Hey! Johnny, okay, okay! I'll sell my car! I'll get you the money tomorrow!"

Johnny Plentenos grunted. He had a thin mustache above thick lips and he scratched this a moment as he regarded the car. "Sell this nice American car, huh?" He spat onto the gravel and dirt parking lot, pulled out a tire iron from the back of his belt and started to bash in the left back fender. The taillight exploded into a hundred pieces.

"What are you doing!" Jake cried. "What the hell are you *doing*?"

Plentenos turned back to Camden, his face a mask of anger. "Don't give me thees *bullshit*, man! You don't own this car, the bank does! You're way behind on payments—no way you sell this car, man! They want to take it back as it is!" The dark man stepped forward, brandishing the crowbar under Jake's nose. "I wonder, Jake. Is it worth fifteen grand to me to put your sorry ass under ground. To no more hear your whiny, lying voice in my ear, huh?"

A fine spray of spittle splattered over Jake's forehead. That, and the sheer force of Pentenos's threat made him cringe back like an oyster doused with lemon juice. "Christ, Johnny, that wouldn't be smart. You know I'm good for the money. Jesus, don't *kill* me!"

He backed away and was about to turn and run, when the two henchmen darted forward and caught him around his arms.

"What you think, guys!" Johnny capered around to face his captive, his big white teeth shining through a grin. "You see, Miquel and Paco—they're my new boys. I get them from Miami, Jake. I'm spreading down to Orlando; maybe I go to Tampa, too, huh? I need some regular muscle, yes. I want to try my new boys out. And their specialty is killing people. Yes, Miquel? Yes, Paco? Or maybe you want Thanksgiving dinner early with this turkey. You grab his legs and make a wish, *sí*?"

The next thing Jake knew, the world went topsy turvy. He was picked up, carried a distance—and suddenly found himself being held upside down over the edge of canal. His head and hands pounded once against the wood siding, and he got a noseful of the creosote coating on the posts, the tar covering the wood.

"Cut it out, Johnny! For God's sake, man, your money's coming in installments!" Camden windmilled his arms trying to get a hold on a post, but the Latin goons, seeing his intention, were somehow able to hold him farther out over the water and mud and reeds. The quarters he'd swiped from the coffee corner at work began to fall out of his pockets, splattering the reflection of the half-moon in the stagnant water below.

"So I see, Jake, so I see!" Plentenos's accent was suddenly gone. He sounded like a normal Floridian businessman; cool, calm, and collected. "I just wish to present you with an ultimatum. You get me the rest of that money by Wednesday of next week, or you are in serious trouble. I just wanted to prove to you tonight that I have the power to do this."

"Yeah. That gives me a week. I'll do it, Johnny, I swear. Just don't hurt me."

Plentenos laughed. "Say, I hear there's an alligator that hangs out around this bar, huh? What do you think, Miquel . . . Is this good alligator bait?"

"We must bloody him a bit, then he make good bait, *señor*," said one of the henchmen, dead-seriously.

"Or cut him up, *sí*?" said the other.

They all laughed. One of the men lost his grip and Jake slipped down a foot before the other man caught hold with his other hand. Jake slammed against the wood, and got a snootful of the foul water and weeds smell from below.

"Okay, fellows. Pull the worm up."

Jake was hauled up and beached unceremoniously on the gravel. He tried to get up but the two Latin hoods knocked him down. One put a large shoe on his chest so close to his chest Jake could smell shoe wax.

Johnny Plentenos crouched down. "Now, Jake. Your ears clean?" Plentenos boxed Camden's ears. Camden swallowed a yelp; they weren't going to kill or maim him, it seemed, so he didn't want to attract any undue attention that might change their minds. "I tell you again. Next Wednesday. I'll be at my club, at three o'clock in the afternoon. I play golf on Wednesdays, when I can. I want it nice, yuppie briefcase. And I want you to walk into the Winston Springs Country Club bar and lounge at 3 P.M. I don't see you, we come and get you. I don't care where you are, Jake. We find you.

Plentenos nodded to the man he'd called Miquel, and the foot got lifted off. Jake started getting up, relief flooding him. But suddenly, a hard toe came out of nowhere, connecting with his solar plexus. The air gushed from his mouth; a handful of stars and a clutch of blackness later, Jake found himself sprawled back on the ground, gasping and wheezing.

Miquel and Paco closed in and delivered a couple more kicks at Jake's chest and abdomen. Sharp tremors of pain rocked him and he tried to crawl away, but fell down, defeated. Suddenly, he found himself in a paroxysm of sickness. The Latin hoods jumped out of the way to avoid the gushing vomit.

Plentenos, already walking away, chuckled. "We kick the shit out of you now, Jake. You don't bring that money, we *kill* you. C'mon boys. We go wipe our shoes off, *sí*?"

The men stared down at Jake, looking reluctant to go just when they were starting to enjoy themselves. Jake tried to say something to encourage them to leave, but all that came out was another stinging gush of puke.

Breathing heavily, feeling as though he'd been burst at the seams, he watched as the bastards sauntered away past the streetlight, their metal-reinforced shoes clicking away like tap night at the Apollo.

Jake got to his feet and tilted toward his car. He wanted to get out of here before those assholes changed their minds and came back. He paused at his car door for a breath, then looked up at the night sky. Despite the streetlamp, the long stretch of canal wasn't lit, and above the magnolia bushes a clear, cloudless swatch of sky glittered dimly with stars strong enough to pierce the moonglow.

"Damn," he sighed, wiping his mouth and staring up at the night. Jake had never had any particular interest in astronomy, although his studies in UFO-ology had necessitated learning something more than he'd gleaned from the SF novels and eighth-grade science classes of his youth. Still and all, he'd always been absolutely lousy with constellations; now, for instance, he couldn't make out a fucking one, just when it seemed very important to do so.

"Damn," he said again, as the pain slowly ebbed from him, a cold, numb fear replacing it. "You little dicks up there ever think about coming down in your ships and picking up Jake Camden for an intergalactic tour . . ." He coughed and spat. "Well, now's the time to do it."

He found his keys and got into his car.

CHAPTER EIGHT

Scarborough was in New York by the middle of the next day.

It had been a little chilly the night before, so he chanced a dingy little motel near the Poconos for a few hours' sleep. He'd tried Ed Myers's private line again early in the morning, with no luck.

He had three possible allies in New York City, all of whom he felt he could trust. His literary agent, his publicist, and his editor.

The odds, he thought, were damn good at least one of them could help him find Diane, or at least give him a place to stay awhile until he could hook up with Ed Myers.

At about ten-thirty, he turned off the Major Deegan in Yonkers. Yonkers was more or less one side of the trenches between the beautiful wooded towns and estates of Westchester County and the steel and concrete towers of Manhattan before he hit the No-Man's Land of the Bronx. It was a curious combination of urban and suburban; tree-lined brownstone villages and large estates in sections, burnt-out housing projects in others. Scarborough parked in an anonymous shopping center, found a public phone that worked, and called his agent.

Ten-thirty was a good time to call people in the publishing industry. Editors and publishers tended to get in around ten A.M., but by eleven chances were good they had to go to a meeting or such; editors and publishers always had lots and lots of meetings, or so it seemed to Scarborough. When he started publishing books, half the time he called, they were in meetings, or a lunch, or had left the office early. He learned quickly that if he caught them at about ten-thirty, he could generally get a telephone audience.

His agent, however, turned out to have inconveniently left for his vacation in the Caribbean.

One down, two to go.

His publicist, Abe Novak and speaking-engagement booking agent was in, though. Abe was reliable. Abe he could always get.

"Everett." The tone was dead-serious, not Abe's usual "Glad to hear from you" greeting. "Where are you?"

"Can't say. I'm in trouble, Abe."

"Tell me about it. I've had a couple of visits at my office, Everett. At my home. What the hell is going on?"

"I guess you've seen the stories they've been doing."

"Yes." And hanging on the wire between them was the publicist's question: Did you do it, Scarborough? Did you kill those men? Huh? Are you a traitor? *Did you do it?*

"I hope you don't believe them!" said Scarborough. "The truth is, I've been a dupe for a branch of the CIA for years! They're trying to cover up something, Abe. And they've got Diane and her fiancé."

"I don't know what to believe."

"Wait a minute, I'm calling for some help, Abe, but if you're going to act like Frosty the Snowman, just forget it."

"I can't give you any help. I have a business and, most important, a family to think of."

"Okay, okay, I know that. But Abe, when I crack this, when it blows over, don't you think I'm destined for the biggest speaking tour of the decade! I mean, if you want to spell it out in dollars and cents, pal, I'm going to be worth my weight in diamonds!" It was a desperate thing to say, pushing money in front of Abe, and he felt slightly ashamed, but he didn't know what else to say. "Look, Abe, anyway, I thought you were a friend."

"I thought I knew you a little better, Everett. Those men—they told me a few things I had no idea of. They were legitimate government men. Now who am I supposed to believe?"

"I see. Okay, Abe, if that's how you feel. I suppose I'd better get off. They've probably got your phone tapped, huh? Just in case."

"I don't think so."

"Thanks so *very* much!" He hung up.

He had to steady himself. It was very hard to accept Abe's rejection.

But still, once he got hold of himself, he really couldn't blame the guy. The big guns *were* against him. It took but a twist of information, a splatter of slander, and suddenly Dr. Everett Scarborough's esteemed reputation was topsy-turvy. These Editors and Publishers, whatever-the-hell branch of the CIA they were, held the trump cards in this game. He could well imagine what they'd told Abraham Novak, and Abe was every bit the rationalist and government-sucking patsy that Scarborough had been. But now... The memory of the past weeks crushed down on him, triggered by his publicist's cold shoulder.

A sudden wave of hopelessness passed through him. For a moment, he felt like just lying down on the sidewalk cement here, and giving up the ghost. He felt naked and vulnerable and alone. He had never felt so ghastly, so detached from the business of living. He felt like a rat in an existential laboratory—*angst* and *paranoia* spelled out in flesh and blood.

Desperately, he wanted his wife. To touch her, to hold her... but she was dead. Frantically, he wanted his daughter, Diane. To see her smile, to hear her laugh. As difficult as she was to get along with sometimes, he knew instinctively that she was the best part of him.

He could feel his heart pumping frantically, and he realized he was breathing heavily. Too heavily. He was moments away from a classic panic attack.

He gritted his teeth. He dug his fingernails into his palms until they felt like they were piercing the skin.

Hang on, man, he told himself. *You can't have a nervous breakdown in Yonkers.*

A gust of breeze blew a Snicker's wrapper up against the pants legs of his jeans. A young man walked by, a large slice of pizza folded in one hand, dripping grease on the ground. Scarborough could smell the bakery a couple of doors down, the rich bread scents, the cakes, and the pastries. In the midst of his shuddering agony, he had an abrupt and unaccountable urge for a slice of pecan pie. He grasped at this the way a drowning man grasps at a life preserver; he focused on the memory of fresh crust, the gooey richness of the filling, the crunch of nuts...

"You okay, mister?"

He looked up. An old woman in a grey wool coat with a

scarf wrapped around her head was stopped on the sidewalk, staring at him. She carried a cart with a banged-up wheel, a few groceries tucked away in its bottom.

"Yes. Yes, thank you," said Scarborough, straightening up, his breathing fairly normally now. "Just a little spell. Uhmmm . . . Asthma."

The woman's rheumy old eyes looked at him for one moment, and suddenly Scarborough's blood froze. Was that recognition in her eyes?

"You're that man on TV," she said suddenly and enthusiastically. " 'Billions and billions'! Cosmos! Yes, on Channel 13. I'm a member of Channel 13. I give them thirty dollars a year! Can I have your autograph, mister?"

Sagan? She thought he was *Carl Sagan?* He must look really terrible now! From seemingly nowhere, his pride rose up, stung. He was about to whip out some nasty retort, when he caught himself. "Of course."

Quickly, he signed "Best, Carl Sagan" on her shopping bag.

"Put down 'Billions and billions,' too," said the old woman. "My Harry, he said that for weeks afterwards. Billions and billions. Poor Harry, he died two years ago."

Rapidly, Scarborough scratched out the phrase. "I'm sorry, ma'am. I have to make a phone call. Nice to meet you!"

She trundled away, her axle squeaking. "Keep watching the skies!" she said.

Scarborough did a double take, blinked, and then had to laugh. "That's what got me here in the first place, ma'am," he muttered. Sighing, he turned back to the phone and dug out his quarters. He knew the number for Quigley Publishing Company by heart, and he dialed it quickly.

Operator . . . assistant . . . and then, he was put through to the executive editor.

His editor.

"This is Cindy Clinton."

"Cindy. Cindy, I'm glad you're in. It's Everett Scarborough. I really need to talk to you."

There was a silence, and Scarborough could imagine the shock registering on that pretty, professionally groomed face.

"Ev. Ev! My goodness . . . I'm so glad you called!"

"I take it you've heard the news stories."

"And that trashy thing on *A Current Affair*. Yes! Where are you? Are you okay?" Real concern sounded in her voice, and a wave of relief flooded Scarborough. Yes, yes, he should have known. Cindy was still his friend. He and Cindy had something more than a professional relationship . . . She could be trusted.

He took a deep, grateful breath. "I'm fine, but I can't talk long. We need to meet somewhere, Cindy. Tonight, in town.

"Of course, Ev. Of course, but what's going *on*?"

"Plain and simple, Cindy, I've been the dupe of the government for over twenty years. Looks like I'm going to have to write another book. If I get out of this alive, that is."

"I don't understand, Ev."

"How do the words *conspiracy* and *top level* and *biggest story of the century* sound to you?"

"Sounds like a best-seller. But Ev, I'm worried about you. You didn't do all this stuff they said . . ."

"Of course not, Cindy. They're framing me."

"Great! That sounds just great, Ev." Cindy was in her mid-thirties. She'd just caught the tail-end of the sixties youth rebellion, but seemed to nurture a nostalgia for those days. Beneath those Saks Fifth Avenue designer clothes, Scarborough had always suspected, there lurked the heart of a hippie. Her father was a former U.S. senator who now had a cushy job with the Defense Department. Rebellion, like Diane? "You tell me where, I'll be there."

"You can help me out?"

"I can give you a place to stay, money—you know, I think there's even a royalty check floating about somewhere around here."

"I'll need it in cash."

"Right. No problem. Just tell me where to be. I want to help, Ev. I really do."

"Okay. Seven-thirty tonight, at the newsstand on the platform at the Columbus Circle subway station. Not the IRT line though—I'll be getting off the D train."

"I'll be there, Ev." Her voice grew soft. "Be careful okay? We don't want to lose you."

"Yeah. Have to go now, Cindy. I can't stay on phone lines long, I start getting jittery."

"Understood. Take care, Grumps."

Scarborough laughed as he hung up the phone. Grumps was Cindy's private nickname for him and his skeptical ways. She'd tagged him that not long after she started working with him. Cindy had been just a regular editor. Thanks in part to the success of his books, she'd climbed up to senior editor—and then executive editor. But that wasn't the only reason she was fond of her pet author—Scarborough had always suspected that Cindy had harbored a secret crush on him. Sometimes, he wondered why he'd never taken advantage of that crush after his wife had died. God knew, Cindy had been supportive enough. And he'd taken advantage of enough *other* women. Cindy was certainly *attractive* enough! A little neurotic, maybe...

Nonetheless, the bottom-line reason was their professional relationship. That, and their friendship. He hadn't wanted to jeopardize either with an affair; besides, he didn't know how to be casual enough to accept such an arrangement, although Cindy certainly went out of her way to be available to him.

Well, she had a fiancé now, anyway. Nice guy named Burke—older divorcé, VP at some P.R. firm. It had all worked out well enough, they were still great friends, she was a very good editor—and clearly, at a time like this, he could rely on her.

Scarborough went into the bakery.

Unfortunately, they didn't have any pecan pie today, but how about a nice Danish, sir?

He bought a bag of jelly donuts and a cup of coffee and then went back out to his car.

Now that he was near New York, he had some other things to do before he met Cindy tonight.

CHAPTER NINE

Betty leaned into his office cubicle.

"Jake," she said. "Mr. Kozlowski just called. He'll see you in his office now."

"Thanks, kiddo," said Jake Camden. He popped a couple Tums into his mouth and chewed them, hoping they would allay the storm that was brewing in his stomach. He slapped on some aftershave and sprayed his underarms with Right Guard. No way did he want to offend the boss this morning, in any fashion. Old Koz held his life in his hands. Sighing, he got up, and put a freshly ironed white linen jacket over his best Hawaiian shirt. His slacks even had creases today. If you could have shined tennis shoes, his Reeboks would have been shined. Too bad his penny loafers had gotten themselves lost in his Fibber McGee closet. Nonetheless, it was a fairly clean and sober Jake Camden that was going to meet Kozlowski this morning—maybe not the Mr. Neat that Diane Scarborough had tried to make him into, but sufficiently changed to make the Old Fart sit up and take notice. Which had certainly not been easy after last night's debate.

Kozlowski was on the phone, leaning back in his huge swivel chair, looking very much the image of a kingpin lording it over his empire.

"Okay, Joe, okay. Yeah, yeah, we still got lunch, we can talk about it then. Camden just walked in. Big story, he says. Right, see you at one-thirty at the Lido." The fat man dropped the phone into the cradle and gave Camden a cursory and contemptuous look as he reached over to a humidor for a fresh cigar. "You want one, chum?"

Kozlowski seldom gave out one of his expensive cigars to visitors, but when he offered you one, you'd better take it and *enjoy* it or he'd be insulted. Jake Camden closed the door behind him, locked it, and then went over and pulled the long crackly thing from its fellows. It smelled of rich, faintly oily tobacco, and was fresh and crisp beneath his fingers. He accepted the light from his boss, took a puff, and then sat down on the edge of the chair fronting the desk.

"Nice, Mr. K. Real nice."

"This better be good, Camden, or I'm deducting the cost of a *box* of those things from your salary for taking up my time."

"Sure, boss. It's good."

"Something wrong? You don't look too hot, Jake. And why the hell are you so *clean*-looking? Your mother visiting or somethin'?"

"That offer you made to me last night . . . those thousands cash up-front for a really great story. Does it still hold?"

"Sure, Jake," Kozlowski's eyebrows knitted with interest. "And more, if it sells those million extra copies, like I promised. Why, something land in your lap this morning? Or have you been holding out on me?"

Camden coughed at that. He gently and reverently put his cigar in an ashtray and folded his hands, ignoring the last jibe, concentrating on the plain delivery of what he had to say.

"This is very big, Mr. K. And we're going to have to handle it with installments, and we can't go into the full truth because of the sensitive nature of the story."

"Yeah, yeah. Christ, Jake, stop hemming and hawing man, spit it out."

"Dr. Everett Scarborough."

"Huh? What, that guy who went nuts last week, killed some people. We already got a story on that from Peters. You didn't want to do it. Other fish to fry, you said."

"I can't go into the details now. But I have the inside story on that whole business. I was there through many of the events last week. And I'm keeping contact with Scarborough on the run from government authorities."

"Jesus!"

The story held the old man's attention sufficiently to allow his cigar to go out. After Camden was finished, a silence settled down between them, but the old man's eyes glittered like polished agate.

"This," he said, "will sell copies."

"I see it as a running series. From the road, as it were. I want it to be very outlaw journalism . . . Like installments of a serial."

"Yeah, yeah! The people will be lined up at their supermarket doors on Monday morning! Besides that, we'll get all kinds of coverage. I like it, Camden! I like it very much. But will Scarborough agree? You say you're still in contact with him."

"No, he won't like it—but the lead-time will make each story about three weeks' history. Who knows, he could be vindicated by then . . . or he could be killed."

"Yes, yes . . . and the first installment?"

"I have to go to New York. You'll have the first installment when I get back." Camden took a deep breath. This was where things got tricky. "I've got some conditions. First, I get that bonus you promised. Second, I still have the rights to the story. I can do it up for another paper or magazine after the series ends. Third, I retain book and movie rights."

Kozlowski picked his cigar back up and relit. A plume of purplish smoke wafted up. He tapped a wad of ash off and leaned over his desk toward his employee. "Wait a minute, I been thinking. You've been holding out on me since you got back. You had this big story, and you kept mum. Now you're spilling your guts. How come?"

Camden looked down at his shoes. Brown stuff peeked from the edges. He must have stepped in dogshit last night or this morning. He sighed. "I need money. Now. I can't tell you why. I just need it. No shuck, no jive here, sir. No bullshit. When I get back from New York, I need fifteen thousand dollars." He stared into Kozlowski's pinpoint irised eyes with all the sincerity he could muster. "It's a matter of life or death. My life. My death."

Kozlowski puffed on his cigar. He looked away, contemplatively, then turned with a steely gaze back to his reporter. "It's the drugs, isn't it? I told you, when you came down here, stay away from that shit, Jake." He shook his head. "You know where I come from, Jake? I'm from Brooklyn. I guess it's not real big secret I ran with the gangs when I was a kid. No secret

I worked with the mob." He shrugged. "America, land of opportunity. Well, I'm legit now, and I'm proud of it. I got some Sicilians who wouldn't mind me being dead, but as long as I stay off their turf, they leave me alone. I'm a tough man, Jake, 'cause I grew up tough. I worked with tough people, many of them stone killers. But I tell you, this drug business now... The Latins... The blacks... Some of these assholes would give Lucky Luciano or Al Capone nightmares. What the mob does—is doing—is just the seamier side of a free economy. What's happening now, though, is total darkness and chaos. I'm starting to think that Satan himself *hablas español*. So Jake, you in trouble with some of these *putas*?"

Jake turned his eyes down and nodded soberly.

"Okay." The boss put his cigar down and tented his fingers. "I don't want a dead reporter on my hands. If anybody gets to kill you, Jake Camden, it's gonna be *me*. Let's deal."

"You'll give me the money?"

"Fifteen thousand. I'll have it for you, cash, when you get back. And no cut for us from any book deal. All I want now is circulation. You get us those extra readers, you'll make this old pecker stiff as a board with happiness. I got plenty of money, Jake. What I want is to be Number One in the business."

"Thanks. You won't be sorry. I swear it."

"I'm not finished, Jake. You keep your book rights, but I want exclusive information—and all of it. You hold out one significant detail of this Scarborough story, you fuck me over again, Jake Camden, you weasel, I swear to God, my alligators are gonna snack on your ass."

"But Koz—this whole thing isn't really *Intruder* stuff. I mean, it could be real, real complicated. I don't think some of it is appropriate to cut up into little bite-size digestible pieces." For a sinking moment, Jake saw his *New York Times* story sprouting wings and flying away. No way was the *Times* going to be printing facts recycled from old *Intruder* stories.

"I'll be the judge of that, Camden. We'll probably edit the hell out of your reports. That's not my point. Just let *me* be the judge of what gets printed. Understood?"

Oh. The old man just wanted to pick and choose. Fair enough. The *Times* story did a loop-de-loop, fluttered back onto his shoulder, and was chirping "Pulitzer Prize!" happily into his ear.

"Understood, sir!" He leaned over the table and pumped his bosses hand. "You won't be sorry sir. And you really have saved my life."

"One more thing, Jake. You hooked on that fucking white stuff?"

"No sir."

"Good. I don't want you fucking with it any more, understood? I get wind of that, you're out of here."

Good enough. Jake could take or leave coke.

"And what about the booze, Jake?"

Oh oh. Jake wasn't so sure he could quit drinking, especially in stressful times like these.

"Well, uh—"

Kozlowski grinned. "A good newspaperman stops drinking, Jake, he sobers up enough to get the hell out of the crazy business. Fuck this New Temperance stuff, huh? Our health editor loves it, and we gotta print the stories for our public image, but me . . . Reading this shit makes me need a drink." Kozlowski pulled out a bottle of George Dickel whiskey and two glasses from his desk drawer. "What's say we celebrate, Jake, eh?" He slopped some of the rich amber stuff into the glasses and handed one to his reporter.

"Why, thanks, sir. I must say, I'm feeling better already. Feeling better about everything.

"To UFOs and to secret government conspiracies," said Kozlowski, standing up and clinking glasses with Camden.

"Yeah." Jake sipped at the whiskey. It was smooth and warm and good. The knot of tension in his gut began to dissolve with the touch of the liquor's heat.

"Who knows . . . maybe we even have some ETs at the bottom of all this! By the way, feel free to speculate all you want, Jake. That's what makes our brand of journalism so wonderful, huh? We're uninhibited by silly facts."

Jake grinned. "You know, Elvis Presley was always interested in secret operations of the government. Maybe the King is running this whole scam, huh? I can see the headlines now: 'Elvis Presley's Saucer Scandal!' "

"Yeah!" said Kozlowski, eyes lighting up. "Try and find a connection, huh? Call up our Elvis editor."

Jake smiled to himself. This was great. Any real facts were

going to get buried deep below the shit. His *New York Times* piece was gonna look brand-new! "The New Tom Wolfe" cooed happily into his ear. "Fame! Fortune! Women!"

"Will do, sir," he said, downing the rest of the drink. "Will do!"

CHAPTER TEN

Everett Scarborough parked his Ford Falcon in a cheap and anonymous lot beside the Hudson River, near West 60th Street. He'd discovered the spot a couple of years ago when he'd made the mistake of bringing his car to Manhattan on one trip and discovered the usurious parking rates in Midtown. Scarborough was not a poor man, but a lifetime of frugality before his book- and speaking-career took off was hard to shake.

It was a park and lock, which Scarborough liked as well, and it was close to the Columbus Circle subway stop. However, he had time to kill. He'd already made several more calls to Ed Myers, calls that met with no success. He'd tried a couple of other avenues of inquiry for Myers via phone, disguising his identity—but he had no luck.

Time to kill . . .

Scarborough wished he'd scheduled the meeting earlier. But as he walked up past Lincoln Center toward Broadway at four-thirty in the afternoon, with the rush-hour traffic and people churning along the concrete canyons of New York City, he realized that he needed to cool his heels awhile. Relax. Everett Scarborough had been running on adrenaline for close to a week. Now that he was back on the East Coast, now that he had built a small but significant network of support, he could use these few hours to take psychological stock. He needed to be in charge emotionally. That seizure back there in

Yonkers . . . He'd never had anything like that happen before, not even during his grief after his wife had died. And yet, there had been something déjà vu-ish about it, something nastily familiar . . . No, there was simply no way that he would be able to help Diane if he was a walking basket-case. He had to get control of himself.

At the corner of 60th and Broadway, he cast about for something to do. Taxis and buses skewed past, kicking up bursts of exhaust and horn honks. Crowds of people pushed past, intent for dinner dates, ballet lessons, or their snug West Side yuppie apartments. He felt safer among this sea of faces than he had on the road, much safer. He wasn't hungry—maybe he could eat at one of the Chinese restaurants along Broadway later. This was one of his rituals during his visits to New York, and instinctively he realized that ritual was what he needed. Up Broadway a bit was a New York branch of Tower Records.

Of course. He'd go shop for CDs. Sure, he didn't have a player in his car, but it was something to do, and since the technological innovation of digital recording had been introduced onto the marketplace, he'd become a real CD-fiend, regularly checking out Washington suppliers to the point where at least one visit a week to a good store downtown was a habit.

Yes, ritual. If anything could help him, ritual would. Besides, who would think to find him in a record shop?

He'd spent much of the afternoon driving the rundown highways and dilapidated bridges which webbed around New York City like a rusty road-graveyard. He knew it was paranoia, but he thought that maybe somebody was following him. Of course, he'd had the trembling beginnings of this sensation ever since he'd struck out from Las Vegas. He'd joked to himself about it after a while; even paranoids have people who were after them! Nonetheless, he couldn't seem to shake the feeling. Now that he was in New York, though, with time in his hands, he had felt that he could engage in some elusive maneuvers.

The New York metropolitan area is particularly tricky for novices to navigate. It consists of a multitude of arteries, veins, and capillaries of traffic, branching out willy-nilly from one another, with dreadful and abrupt signs marking awkwardly angled turnoffs and changes. The numerous bridges and tunnels that connected the island fortress of Manhattan with its neighboring boroughs and New Jersey provided many opportunities for sudden kamikaze lane-changes to shake pursuit. So Everett Scarborough spent more than two hours that afternoon driving about the Big Apple's thoroughfares like a madman. It was fortunate that most New Yorkers drove like crazies as well, or he would have gotten a ticket. If he had thought that driving like an Italian in a parking garage would have jeopardized his freedom, he would not have done it. But hey, he thought. This is New York, New York, where fender-benders were a mode of affection.

He'd actually found that concentrating on this course of action had improved his mood. He felt more in control now. And certainly, if there had been anyone actually following him, they weren't doing so anymore. Most likely, they were tangled in traffic on the Brooklyn Bridge after his daring turnoff onto the FDR.

Anyway, the feeling of being followed was gone.

Funny, he thought as he walked past the hot dog vender by the ABC Studios with "All My Children" lettered on its marquee, and then hurriedly cut across Broadway to take advantage of the Walk sign. He'd been a man who discounted things like intuition and extrasensory perception. Now, it seemed, with his ordered and rational world turned topsy-turvy, he was a sudden, if reluctant, believer.

Tower Records was a large two-storied store brightly lit with neon, strewn with promotional posters, and simply choked with records, tapes, and CDs. The Tower Records in Washington, D.C., which Scarborough normally patronized, had a video section, but in New York there was a separate store up the street marked Tower Video.

He'd always collected records, particularly jazz records, from the late fifties on. He had maybe three or four thousand records, many of them collectors' items worth a lot of money. The advent of compact discs, however, and CD players had been a double blessing. First, they saved his collectors' items from further deterioration due to playing. He owned several copies of Dave Brubeck's "Time Out," for instance—and all but one were extremely scratchy by now. With CDs, you could play them forever, and there was no wear because there was no friction involved—only a laser reading bits of data. Second,

CDs made a midlevel system sound *great*. Oh, maybe the purists screaming in the *Absolute Sound* were right; maybe if you had a $30,000 stereo system with vacuum tubes, you had superior sound. But you still had the snaps and pops and scratches from older records—the brilliant remixing and engineering jobs being done on old jazz records simply weren't being reissued on vinyl format . . . they were only coming out on CD. Besides, there was something sexy and snappy about the shiny disks in their neat, easily filed jewel-boxes.

That, and at least these disks didn't *fly*.

Yes, CDs had put a brand-new sheen on his music collection, and since CDs of old stuff were being released ever so gradually, part of the fun was going out to shops and *looking* for them. There was the thrill of finding that old Charlie Parker collection—or that new anthology of Stan Getz recordings from the fifties, say. A harmless addiction, since Scarborough had the money to pursue it.

After a while, disappointed with the selection here, he decided to give up on jazz and go check out the classical section. Of course, he wasn't actually going to *buy* anything, but just being among these things, panning for gold as it were, connected him to more relaxed and certainly gentler times.

The classical collection here was huge. Cultured people in jackets and ties, with well-trimmed facial hair and the telltale air of CD-fever examined the new-release sections carefully. Down here it was air-conditioned. *Four Seasons* by Vivaldi was playing quietly on discreetly placed BOSE speakers. Calming immediately, Scarborough settled down for the hunt.

There was that collection of Julian Bream guitar interpretations of Bach that he'd lost. Maybe if they had it down here, he might even shell out thirteen or fourteen dollars to pick it up. Stupid, maybe, since he didn't have access to a CD player, but the very act of tendering the plastic-wrapped treasure, along with money, was an inevitable part of the CD shopping ritual.

He found the Bream-Bach CD and felt the familiar thrill of discovery.

It was then, when he looked up unexpectedly, that he noticed the two men.

They looked much like most of the other men in the store; generally office workers on their way home from work, stopping in to browse. Pinstripes, ties. One looked to be in his late twenties, early thirties; the other closer to Scarborough's age, with salt-and-pepper hair. They had also been up in the jazz section, over in big-band territory. Now they were down here, and what was more, Scarborough had the distinct feeling that he'd seen them somewhere before. Not here in the record shop, not here in Manhattan, but *somewhere else*.

But then one went off to look at Mozart disks, while the other wandered over to the soundtrack section.

Scarborough continued his idle searching, again shaking the notion that these guys were following him.

Anyway, who could they be? If they were CIA the FBI or plainclothes policemen, and they knew who he was, why didn't they just come up, arrest him, and cart him off? It was pretty clear that Scarborough was unarmed—he wore jeans and shirtsleeves today. Besides, there was no way anyone could have followed him. He'd well and truly weaved a tangled web, and nobody but Marsha and Cindy knew he was anywhere near Manhattan.

No, he decided. Just his jumpy paranoia, hopping up again.

He forced himself to look for ten more minutes, then went up and bought his Bach CD.

He wandered up Broadway to find that Chinese restaurant.

It was a rested and well-fed Everett Scarborough that stepped off the D train into the Columbus Circle subway station. He'd had to walk a ways uptown to catch the D, but that was what he'd promised Cindy, and that was what he had taken.

When he got off the train, he walked over to the newspaper stand, his eyes sweeping over the crowd. For some reason, the station was more crowded at seven-thirty than he remembered it being from previous visits. Still, it was far from rush-hour level. At the newsstand, however, he saw no sign of Cindy. Nervously, he looked up at the clock, and noted that it was only 7:25. He was early, then. Well, better than later, he supposed.

His back to a green-painted, steel-girder brace, he waited for his appointment, watching the people walking through the station, leaving on the roar of subway trains, arriving in a humid rush of moving bodies.

Every time he came to New York, two things always surprised him. One, it was even more expensive then he remembered, or even expected it to be. Two, it was dirtier, with a peculiar grey squalor unique to this place, touted as the greatest city in the world.

He'd always heard that you had to have been raised in the New York environs to be truly inured to the sandpaper-on-nerves, in-your-face-and-up-your-nose sensory overload it confronted you with every single waking moment (and who knew, maybe *sleeping* as well). Scarborough was just very happy he didn't have to live there, only had to visit once in a while.

Offensive as the topside was, it was the subways of New York City that always shocked him. They were clangorous horrors, dark and dingy, smelling of old and fresh urine when they didn't smell of body odor and rat droppings. They were horror dungeons, connected by the dank tunnels of Hades, illuminated fitfully by splashes of electrical sparks and seemingly phosphorescent ghost stations. Scarborough, when he rode them, half-expected to gaze out into a scene from the twisted landscape of Hell, and find it a marked improvement to the war bunker coming up on 14th Street, or Tenth Avenue, or wherever.

The Columbus Circle Station, in which he stood, waiting, was typical. Upstairs would be the usual bums and homeless people, defecating in dark corners and begging for spare change. "Reagan's people," Scarborough called them, and while D.C. more than had its share, thanks to Ronnie R.'s Nazi-like programs, the species seemed particularly virulent and lice-ridden here in New York. Scarborough always gave them money. He couldn't help himself. Friends here told him, just don't make eye contact. But he couldn't help looking, feeling a surge of pity and fear. One of the nice things about being well-off financially was that you could make contributions to charities for the unfortunates like these, but no amount of charity could hide the scars of the mismanagement of the social mechanism that should prevent this kind of thing.

Yes, the Columbus Circle subway station was filled with the newspaper-shod homeless and it smelled; its ceilings were low and the girders were ugly and it was filled with trash. It was a study in Industrial-Age concrete and metal gone wrong, through which screaming worm-monsters rolled ceaselessly through their haunts of the dying city.

Despise the subway system of New York as he did, however, Scarborough always had to go down and ride it. Why, he didn't know, he just did.

He figured that the Columbus Circle station was a good place to meet Cindy for three reasons: one, it was easy to get to for both of them; two, it was public, and he could melt into the crowd; three, it was easy to get away from, with numerous avenues of escape.

At exactly 7:36, he saw her coming down the stairs from the ticket-buying level, looking around for him. He took a moment to check the area before he waved her over, to make sure that no one was following. But what did he expect? he asked himself. Men in trench coats who looked like Robert Ludlum. He made himself relax, stepped out from behind the steel girder and waved.

She saw him, smiled, waved back and continued down the stairs toward him.

She was an attractive woman with styled and highly moussed auburn hair and round aquamarine designer glasses with their frames cut low for better peripheral vision. She had an oval face, freckles from a recent Caribbean vacation unhidden by makeup. She wore a black business suit, low pumps, and carried a large black bag slung over her shoulder.

Scarborough viewed Cindy with a curious ambivalence. On one hand, here was a bright and attractive, highly ambitious young lady, an excellent editor, with excellent taste in clothing, painting, interior decoration, food, what have you. She also had excellent taste in men. It had been clear from the very beginning that she fancied Scarborough—this was even before his wife's death. This schoolgirl crush had mellowed out as she'd matured, into an uneasy friendship below the professional relationship, and then into more cynical lust that they both joked about, once Scarborough decided that he didn't want to get physically involved with her. It will jeopardize our editor-author relationship, he had told her, but it was more than that. There were things about Cindy that bothered him, certain areas of coldness and neurosis that showed through from time to time, like chunks of icebergs in the Tropics. Chalk it up to normal New York psychological disorders, he'd thought. This wretched city did that to women. Oh, he'd had his share of

New York women, and they generally complained at the lack of straight, unmarried potential partners in this town. Anyway, he'd steered clear of ending up in the sack with Cindy, even though there had been a few close calls at champagne-fueled ABA- and book-parties.

"Ev," she said now, walking up to him with crisp, confident strides, her breasts jouncing beneath the crisp white silk blouse, her sleek smooth legs scissoring along, motored by those nicely tailored, nicely proportioned hips. "Hi. There you are. I'm so glad you're safe."

She kissed his cheek and he got a whiff of White Linen perfume, heaven amidst the olfactory horrors of the subway train station.

"Thanks for coming, Cindy. You didn't see anyone following you, did you?"

An odd expression passed over the rounded angles of her face. "No. Should I?" Her accent was Ivy League smooth, with excellent diction, a perfect complement to her bright but troubled eyes.

"I just can't be too careful, you know."

"I understand, Ev. With all this craziness. Good disguise, by the way. I'd hardly have recognized Everett Scarborough, poorly dressed." She gave a little flicker of a smile. "So. Where to from here? Do I get the honor of having my most popular author spend an illicit evening in the boudoir of his editor?"

He had to smile at the tart joke. "Actually, that's not what I had in mind. They might trace me to you and figure out I might stay at your place. Show up at three in the morning. Catch us flagrante delicto."

"That's unlikely, with our history." But the joke was through a frown, and he got a sudden strange feeling. It started at the base of his spine and moved up like a icicle on centipedal legs. Something was *wrong* here. Very wrong. It wasn't that Cindy was acting strange. It was just the opposite—she was acting *normal*. No tics, no neurotic fidgetings she occasionally had when under pressure. Here she was, meeting with a wanted criminal, and she was acting like she was meeting him for drinks and a flick.

"Look, Cindy, I need traveling money. Did you get that advance?"

"Sure, but I'm not carrying it around with me. I'm afraid you'll have to brave the jungles of West End Avenue for that, Ev." He sensed a forced friendliness. A brittleness.

"I really appreciate it, Cindy. I can't be too careful. I'll call you tomorrow morning and make arrangements to pick it up somewhere, all right?"

"Everett, are you okay? You're acting very funny."

"I don't know, Cindy. I just don't know. You're just going to have to go along with me on this. Go on home now, or wherever. I'll be in touch."

He turned away.

GET AWAY FROM HERE! something screamed inside of him. *GET AWAY FROM HER!*

"Everett!"

He started walked toward the exit. The southbound A-train was barrelling into the station, covered with graffiti.

"Everett Scarborough, stop."

The voice was closer. She was behind him, keeping pace.

"Sorry, Cindy. I have to go."

The doors of the train opened, disgorging passengers, taking them in.

"Scarborough." Her voice was hard now, tough and determined. "I have a gun on your back. Stop, or I'm going to shoot you."

Surprised, Scarborough wheeled around.

Cindy stopped; one hand supported her open purse, the other was shoved inside it. There was something hard pushing against the black material. The gun? The forefinger-in-the-pocket trick? Whatever it was, Scarborough was stopped well enough by the plain obviousness of the fact that this woman was *not* on his side.

"Cindy," he said, his insides tightening as though someone had punched him in the solar plexus. "Cindy, what's going on?"

"I'm taking you in, Everett. For your good, for everyone's good. You've become too much the loose cannon."

She was looking around, and Scarborough saw who she was looking for before she did. A man in a grey suit and short hair, every fiber of him spelling FBI was walking down the stairs, with another pair of grey legs behind him.

She followed his gaze. He took his one chance of getting out of the situation, and he jumped forward, knocking her purse and hands to one side and clipping her a hard blow across her face. She went down without firing. He grabbed her, picking her off the filthy floor, looking up at the men.

They were starting down the stairway, their hands reaching past the buttons of their jackets.

One hope. Scarborough, adrenalized by the danger, found that Cindy seemed light as a feather. He picked her up, and dragged her into the nearby car of the C train, just the split-second before the doors closed. The train started up, and was rolling along quickly by the time the two men in grey pounded up beside it, looking angry and helpless. Not taking any chances, Scarborough shied away from the windows, pulling the stunned editor with him.

"Let me *go*!" she screeched suddenly. "Get *away* from me!"

The agents (FBI? CIA? What difference did it make? They were *after* him, that's all he knew.) slammed against the door, screaming at the motorman to stop the train. But the screeching of the wheels and sounds of the station swallowed up their orders. The subway train soon slithered into the dark tunnels, away from them.

The passengers of the sparsely populated train moved away from the two as though they had the plague.

"Help me!" cried Cindy. "He's a *criminal*!"

"Sorry about my wife," said Scarborough, smiling over to the motley collection of passengers. "She's distraught about this affair I'm having with my secretary."

"There's a policeman on this train! Call him!" cried Cindy.

A man near the exit pulled the door open and hurried off to the next car.

Yes, maybe there was a cop on the train, but it would take him awhile to get here. They couldn't stay on the train, of course. Not for long, anyway. But as long as he had her here, hand gripped hard on her forearm so she couldn't run away, Everett Scarborough had some questions for the executive editor of Quigley Books.

"Who were those men back there?"

"What men?"

"Your backup. FBI? CIA? What's going on, Cindy? Did they have your phone tapped? Did they force you to do this?"

"That's right. They made me, Everett."

He looked at her, but although fear shone in those blue eyes of hers, sincerity did not.

"Who *are* you?"

"Cindy Clinton, Everett. They knew you'd contact me, so they made me try and bring you in."

"No, who are you *really*?" That feeling again, deep in his gut. He realized that it had all been just an act for her, from the very beginning.

"I don't understand."

It came to him in a flash. A realization like lightning, ripping his ego apart. No, no, it couldn't be . . . But maybe it was . . . It had been Cindy who had bought his first book when all the other publishers had turned it down. It had been Cindy who had worked with him, chapter by chapter, promotion by promotion. She had risen as *he* had risen, she claimed.

Funny though. She'd always had a nice apartment. She'd always had nice clothes, a car . . . And those exotic vacations. Scarborough had always wondered about those. A little family money, she had always claimed when he'd inquired. But that wasn't the truth . . . No, not the whole truth, he realized. She'd come down to Washington a hell of a lot come to think of it . . . more than necessary just to work with an author. To see daddy she'd claimed. He'd been so blind. The *connections* were so plain, so clear. But then, in the ecstasy of success, he'd never noticed the strings they'd been planting in him even then. *Especially* then!

"Tell it to me, Cindy! Tell the *truth*!"

"Stop! You're hurting me."

"You don't just work for Quigley, do you? You work for the Publishers and Editors!"

A little jar of shock in her eyes (how did he know those names?), fading quick as she got hold of herself again.

"Publishers and editors. Well of course I work for publishers and editors. That's my career."

"Don't play dumb. I mean, the *CIA*. A shady part of . . . Maybe even more than the CIA. I don't know . . . but you're going to tell me!"

"How can I tell you what I don't know. They're right, Everett. You have gone over the edge."

"No, I'm just wising up, Cindy."

"They don't want to hurt you, Everett. They *want* to help you. Come back to my place. We'll arrange for a peaceful conference. Them and you, that's all they want."

"No, Cindy. I want to talk about *you*! I want to talk about the lofty publishing career of the bestselling skeptic-scientist in the U.S.A. They were in on it from the very beginning, weren't they? They had you infiltrate a respectable publishing company, acquire my first books . . . And then they built the machinery of promotion and fame to make me a star. Didn't they, Cindy! *Didn't they*!"

"Everett, let go of me. My arm's going to fall off."

He realized that he'd been holding onto her much too tightly, and he let go. He relaxed a bit, but still kept a firm grip. The train rolled into a stop at 50th Street station. Fortunately, no men in grey suits got into the car, but a number of the more respectable passengers took the opportunity to get off or change cars. Scarborough felt certain no one would go for a cop—he and Cindy had settled down to what looked like a domestic squabble.

The doors rattled closed. The train rumbled down the line.

"I don't know what you're talking about, Everett, but please, trust me. I'll get you help. There's something wrong with you!"

"No. There's nothing wrong with me. I see it all now. I should have guessed. They had me coming and going. I was their little bright and chipper, suave-scientist *patsy*. It was so easy—I didn't even have to sell out—just build up a gigantic spoon-fed ego!"

"You're paranoid!"

"Oh yeah? Then how come it was necessary for you to pull a gun on me, Cindy? How come it was necessary to bring out those grey suits?" He took a deep breath, gritted his teeth. "You're coming with me, sweetheart. We've got some serious talking to do; oh yes, we do."

"About what? Everett, this is kidnapping. You're just getting yourself into more trouble."

"It's your people who're the kidnappers. Where are they keeping Diane!"

"I don't know what you're talking about!"

"They—no, *you* and your cohorts have got my daughter, dammit! What have you done with her!"

"Everett, please. Look, okay, I admit it." She looked away, unable to face him. "I am a government official. I serve my country, and I'm proud to do so. What I'm not proud of is what I've done to you. Do you think it's been *fun*, Everett? Haven't you seen that I've developed *feelings* for you? Don't you think all this has torn me up inside?" She sighed, and Scarborough could see a glistening tear sliding down the side of her face. "I guess I told myself that you'd never find out, that you were *enjoying* yourself so much, with your fame, your success, that it was really *good* for you. I never realized it would come to this." She turned to him, and he could see the sincerity shining through those troubled blue eyes. "I swear, Everett, I'd never, *ever* do anything to hurt you or Diane. I wish you'd just stayed away from me. I had a choice, Everett. Betray you, or betray my country. If you'd just come in with me—well, okay. Everything *won't* be all right. But we can work out some kind of arrangement beneficial to everyone. Right now, Everett, you're a loose cannon on the deck, and these people I work with—they *have* to stop you. Whatever way they can!" The train was slowing down now, coming into Times Square Station with the slow squeal of brakes. What he would do, Scarborough thought, was to go on down to 34th Street at Penn Station, take a train uptown, find his car, and get the hell out of New York. He couldn't force Cindy to come with him; that would be far too dangerous. Some cop would stop him eventually. No, she had to go along of her own accord.

"Tell me about the Publishers and Editors, Cindy. Tell me what they've done to my daughter."

"I can't tell you anymore. I shouldn't have said anything at all, as it is." She was clamming up again, damn her. The tears were gone. She was a trained government agent. The only reason she'd gone soft was because she'd had an emotional attachment to him. But it was clear where her true loyalties lay.

"Cindy, for *God's sake*, what's *wrong* with you? A country that indulges in these kind of operations must have something rotten at the core! They've brainwashed you, Cindy. You're as much a pawn as *I* was. For God's sake, help me! If you want to serve your country, it's the Constitution and the Bill of Rights you have ultimate allegiance to! Not a bunch of hoods and killers out to—"

The train shuddered to a halt, and the doors opened. A big black teenager sauntered in, all Pimp Troll attitude, rolling biceps and a huge boom box blasting rap music. The explosive entrance attracted Scarborough's attention, and he loosened his grip on Cindy . . . forgetting she was a trained government agent.

Her blow took him by surprise, a quick jab across the face. He was knocked back, blinking with shock and pain.

The next thing he realized was that Cindy was running through the open metal and glass doorway, her pumps clacking onto the cement of the 42nd Street station. The station was more crowded here, and she dodged New Yorkers, acting as though she knew exactly where she was going.

He dashed out after her.

"Cindy!" he cried. "No! We have to talk. I swear, I won't hurt you!"

Heads swung away a split-second, then averted into the classic New Yorker's dead, I-don't-care stare.

He ran after her.

But then he saw her destination.

Across the walkway, on the other side of the platform by a grey girder, stood one of the New York's finest, dark hair hanging from behind a casually slung cap, talking into a walkie-talkie. But he was looking the other way, and there was a train approaching. The din was so tremendous, the cop clearly could not hear Cindy—or for that matter, Scarborough yelling out to Cindy.

Scarborough stopped, halted by indecision.

He turned around. He had to jump back on the C train. But even as he swivelled about, he saw the doors closing, the train starting to lurch away.

Damn! He turned back. Was she really heading over to a cop? If so, her intent was clear. She was going to make up some kind of story about assault, and pin it on him. The cop would radio for help and he'd get caught. He had to be arrested only for a short while before the grey suits closed in, and then that would be it.

He had to run!

But even as he had this thought, as he watched Cindy closing in on the cop, raising her hand to grab his shoulder and turn him around, something quite unexpected happened.

Cindy jerked.

She jerked as though some invisible hand had struck her. A split-second later, a little spurt of blood sprayed out from her back, splattering the grimy grey floor of the platform.

Christ, someone had *shot* her!

Her mouth opened, but any scream was swallowed up by the roar of the train coming into the station. Her arms windmilled, and she lost a shoe. She changed angles, splayed wildly, her designer jacket stained crimson with her own blood.

"Cindy!" Scarborough cried.

Jesus, she was headed straight for the gap! She had gained plenty of momentum with her run, and the added impact of the bullet from nowhere pushed her harder. Out of control, she tripped, and hurtled straight out into empty space, and then down onto the rails, face contorted with panic and horror.

There was an immediate screech of brakes as the motorman saw the woman fall onto the tracks, but the train was only a quarter of the way in, and it still had plenty of speed.

It rolled right over her.

"Cindy!"

Scarborough could almost see those hard steel wheels grind over that shapely body, severing limbs, crushing her skull, mincing her up into a gory slab of bone and twisted skin and gristle. He almost felt the bright singe of pain, and then the descent into total, forever blackness . . .

"Oh, Jesus!" he cried. "Cindy!"

People were beginning to scream.

The cop's attention had been attracted; he just stared down at the accident, eyes bugging with shock. It took only two or three seconds for the confusion to swell into chaos as some of the crowd moved away, some froze as still as Scarborough, and some moved forward to rubberneck.

Cindy.

Someone had shot her. But why? Because she was going to turn him in. Because he was in danger?

It didn't add up.

Unless . . . Scarborough was unable to keep the thought out of his mind. And the image came back to him unbidden. . . *Standing in the open, on top of Hoover Dam, taking a wild shot at the assassin who meant to kill him. The impression that the shot had gone wide—but then the impact and the blood as a bullet struck the man, pushing him back and over the fence on the side of the dam . . .*

The train had stopped, and the cop was barking something into the walkie-talkie. Scarborough found himself drifting over toward the side, hoping against hope that somehow Cindy was still alive, that she'd wedged herself against the wall, avoided the rending wheels, the crushing weight of the train.

A hand clamped onto his shoulder. Oh Christ, the thought flooded him. They've got me.

He was spun around powerfully, and was face-to-face with a man in a suit and tie. Later, Scarborough remembered that the man had a peculiar smell. A smell that, at these close quarters, made his flesh crawl with goose-pimples. He was, however, far too juiced with adrenaline to really notice at that exact time.

"Everett Scarborough," said the man, regarding him with an implacable stare. He spoke with almost perfect diction, in a deep tone of authority, like a radio announcer. "You must leave here. Now. Or you will surely be captured."

"Who are you?" Scarborough said.

"Leave, or we shall have to *kill* you, Everett Scarborough!" A note of worry. Excitement? The man's hand reached into his jacket, showed the edge of an automatic gun. "That would be a tragedy. A true tragedy."

"You! *You* killed Cindy!"

He recognized the man now. He was one of the pair in Tower Records, flipping through the classical records.

A great commotion sounded from the top of the platform. Scarborough saw the uniforms of rescue workers and policemen.

When he turned around again, the man dressed in grey had disappeared, swallowed up by the crowd. Scarborough felt a sympathetic pain in his chest, almost anticipatory of a bullet smashing through.

They were out there, watching him, those two men.

But how much longer did he have before the people who wanted to capture him, *wanted* to kill him.

Confusion raged through him, confusion and grief. But now, survival had again nudged him. He needed to go unarrested, free, if he wanted to solve this terrible puzzle, if he wanted to find his precious daughter.

He turned away from the chaos and walked slowly up the steps, away from this scene of sudden and senseless death.

CHAPTER ELEVEN

Maximillian Schroeder answered the door himself.

The dough the guy had these days, Jake half-expected a hoity-toity butler to answer the door. But no, here was Max Schroeder in the flesh, gabardine slacks, polo sweater, and Weejun penny-loafers. He peered out at Camden through hornrimmed glasses, his curly hair disheveled in a tasteful Thirtysomething style.

"Max!" said Camden, grabbing the man's hand and pumping it enthusiastically. "Max, good to see you, pal. Thanks for lettin' me come see you!"

The hand was smooth and limp. Schroeder allowed Camden to shake it for just two seconds, then pulled away. "Jake Camden. Please, come in, Jake. I've prepared some breakfast. Would you like some coffee?"

Jake grinned. "That's awfully kind of you, Max. As I just pulled myself out of bed mere moments ago, I suppose a cup of coffee would do real well now, and maybe then I can choke down some toast or something."

Schroeder shook his head sadly and closed the door behind

his guest. "Breakfast is the most important meal of the day, Jake. You should eat a balanced one."

Maximillian Schroeder lived in a brownstone on 63rd Street, on the East Side of Manhattan. He and his family had the entire brownstone to themselves. Such a building was worth millions, and though Maximillian Schroeder was a rich man from his books, he was not perhaps *that* rich. The brownstone was part of an inheritance from a wealthy uncle who'd made his money in import-exports. Schroeder's family in Oklahoma had a good deal of money as well, which had brought him his Ivy League education and his general air of intelligent erudition. Schroeder was also a highly successful novelist. But it was his nonfiction books that had made him a mint. His books about his experiences with UFOs and the beings who peopled them. His books about his abduction, and his subsequent communications with the "Others," as he called them. His books that some found frightening, and others inspirational.

"Never got in the habit. Maybe a little jelly on a bagel. Mind if I smoke?"

"This is a smokeless environment, Jake."

"Yeah. I kinda figured." Jake put his pack of Camels back in his pocket. He really didn't want a smoke badly, he was just testing the boundaries. "Nice place you got here, Max." Jake looked around as he was escorted through corridor past a living room, back to the dining room. Everything was decorated in the best of old-fashioned taste. The rooms had an antique and expensive air to them, but fashionably lived-in, as befitting a genteel author with decidedly bohemian and bourgeois fellows and audience. However, the kitchen, which Camden could see beyond the mahogany, teak, and walnut of the dining room, was decidedly yuppie high-tech, with chrome sparkling and tile gleaming.

"Thank you." Schroeder gestured for Camden to take a seat. The breakfast on the side table consisted of coffee in an immaculate silver service, a ceramic urn's worth of oatmeal, skim milk, brown sugar, and a bewildering variety of fresh fruit. Lying on the table was a copy of today's *New York Times*, which Schroeder had clearly been reading when Camden arrived. "I really prefer our more rustic quarters out in Pennsylvania, but Melinda likes to be close to her cultural events.

Lincoln Center, Carnegie Hall. So we spend a few weeks here, a few weeks there.'' Schroeder's Oklahoma accent was minimal, but it occasionally glimmer through, complementing his soft-spoken manners.

''Pennsylvania. That's where you met the ETs, huh?''

Schroeder grimaced. ''Please, Jake. The 'Others.' We're not living in a Spielberg movie.''

''Oh, yeah, sorry. You know, I keep on forgetting that you take all this stuff really seriously.''

Schroeder took off his glasses and leaned toward Camden, his steely blue-grey eyes dark above a frown that wrinkled his narrow face. ''It is *serious*. We are at the dawn of not only a new age, Jake. We live on the cusp of things beyond our imagining. There are things the Others tell me which are far, far beyond my comprehension. My mind burns at times, Jake, *burns* with the implications of the merely metaphysical nuances of the messages the Others present me.''

The man's gaze was so sincere, so penetrating that Camden had to look away. He drank some of his black coffee, and cleared his throat. ''Yeah. Maybe so, Max. Anyway, I need some help from you, some advice. Maybe even a quote when I get around to writing this book.''

''Another book, Jake? I hope it has a good deal more credibility than the last. Ninety-five percent of what is written on the subject of UFOs and extraterrestrial life is so unfortunately muddled. Shadows on Plato's cave, Jake—particularly distorted shadows. I could not possibly endorse a book of yours unless I was sure it was sincere—and of course, I shall have to correlate it with the Others.''

''Sheesh, Max. Get the broom out of your ass and give me a listen, huh? I've got some heavy stuff here.''

Schroeder's eyebrows raised with extreme interest. ''Have you, now?''

''Yep. Heavy-duty, hot news, and I'm dead in the middle of it. When this appears in the *New York Times* or maybe the *Washington Post*, I'm gonna have credibility up the old wazoo!''

''How nice for you, Jake. So what is it this time? Cattle mutilations by the Slime People of Venus? Elvis impersonators from Alpha Centauri?'' A condescending smile appeared on those famous thin, wide lips.

"No, Max." Camden leaned toward him, grinning with a victory that he really hadn't had the opportunity to celebrate until now. "The true story of Doctor Everett Scarborough!"

Schroeder stopped smiling. He blinked, taken aback. "Scarborough?"

"Yeah. And guess what. He and I are *buddies* now!"

Maximillian Schroeder recovered rapidly. He leaned back in his chair, picked up his coffee and sipped it thoughtfully. "Doctor Scarborough is a brilliant man. I respect him greatly. However, I thought that he openly despised you, Jake."

Camden grinned. "Pure past-tense now, pal. Those days are gone." Camden held up intertwined index and forefingers. "We're like that."

This bit of news clearly was taking Schroeder awhile to properly digest. And no wonder. He and Dr. Everett Scarborough were the greatest of gentlemen enemies. Each considered the other his own private Moriarity. Schroeder, of course, was the more gracious of the two when they met at UFO functions or, most especially, on television shows. However, it was always clear that Scarborough's sarcasm, his scathing wit, his command not only of science but of language, wounded Schroeder deeply. He was a man who hungered for dignity, respect, perhaps even acclaim, despite the wild and wacky message buried beneath the acreage of soft-spoken sincerity.

Maximillian Schroeder was famous for his obliging politeness to even the most off-the-wall saucer freak. Part was his personality, perhaps; but mostly, it was part of the Message he preached, as received from the Others: Everyone on Earth is communicated to by extraterrestrials in one form or another, but as each person is different, and each message processed differently, most are squelched by the engines of the subconscious, and the ones that emerge at all ray out like light beams shot through the multitiered prism that congeries humanity.

Maximillian Schroeder had not started out a Prophet of the Star People. Hardly. He'd started with a terrific education, a fortune, and, starting in his late twenties, a successful career as a writer of thrillers. This began with the famous Codename books which regularly did well both in hardcover and paperback. *Codename: Werewolf* was the fist. Then came *Codename: Vampire*; *Codename: Psychic*; and finally *Codename: Conspir-*

acy. These were followed by a couple of generic thrillers that did not do as well. Then Schroeder suddenly revealed to the entire world (through a book named *Baptism*) that not only had aliens from other planets kidnapped him and taken him aboard their flying saucer, but he realized that this sort of thing had been happening to him for a long time. *Baptism* had sold millions of copies worldwide. His subsequent books, *Touched* and *Pentecost*, had done almost as well; and when he wrote a novel called *Revelations*, about the people he called the "Others" —dictated, he claimed, by a mass-mind communion of his own—the book went through the roof of the bestseller lists.

It helped perhaps that it was now the 1990s, close to the zero-hour of A.D. 2000. Millennial paranoia had fueled the interest in the book, since it was the story of how beings from another planet landed to bring peace and prosperity to mankind after a terrible conflict with fascist governments and naysayers. The credulous called Schroeder a "visionary"; his detractors called him an "opportunistic rabble-rouser." In actuality, as far as Camden could judge from his experience with the man, Maximillian Schroeder was a painfully earnest individual with some severe psychological dysfunctions that somehow had not turned him into a psycho or sociopath but had led to fame and increased fortune instead. He was a real nice guy, if a little weird at times and Camden truly liked him, if for no other reason than he put up with Camden's bullshit.

"I don't understand. I thought that Scarborough was in a lot of trouble with the authorities," said Schroeder. "How could you get involved?"

"Genius, pure genius." Camden briefly outlined his involvement, careful not to give any really important details. "The poor guy is as framed as the Mona Lisa, Max."

Schroeder shook his head. "I suppose I'm feeling a little sorry for Scarborough, but frankly, what does all this have to do with me?"

Camden laughed. "Not a whole hell of a lot, actually. I just thought it wouldn't hurt to let you in on some of the details. There's some kind of heavy-duty conspiracy going down in the government—I mean, *bigger* than the MJ-12 documents, bigger than just about anything that's come down the UFO pike."

"But Jake—flying saucer enthusiasts have always assumed

that there's been some sort of U.S. government cover-up. This cover-up, indeed, has apparently spread throughout the world. Every UFO group in every world claims that authorities suppress the true evidence. This is a tradition."

"Come on, Max. You've been screaming along with all the rest of them. I've heard you talk on the subject!"

"True."

"Then you *do* think there's a conspiracy going down."

"Oh, certainly. But I don't think it's all that important. When the Others wish to make themselves fully known to the world at large, there is absolutely nothing that anyone will be able to do to stop them."

"Others shmothers, Max. Let's get down to the nitty-gritty, away from the aerie-faerie stuff. You pay your taxes, you have a right to know why the CIA, the Air Force, the NSA, and what-the-hell-have-you is preventing the message from going out. Maybe your aliens *have* sent a message to the world at large—and the boys in D.C. are stonewalling it."

"I shall make a sizable donation to the Scarborough defense fund when the man is apprehended. Is that what you're collecting for?"

"No. No, Max, I'm just trying to pay you off with some interesting information for the stuff you're going to give me?"

Schroeder shook his head, a little baffled. "I'm sorry, Jake. But I don't understand what you're talking about."

"Don't you see what I have on my hands, Max? I've got the greatest story of the decade . . ."

"Which will be printed in newspapers. So, what do you want, a little touch of vitriol from me? 'I always knew that, deep down, Scarborough was nothing more than an outlaw!' " said Maximillian with a satisfied "Tsk, tsk."

"No, Max. There's going to be book offers! There's going to be a possibility for movie offers—based on the book by yours truly. We're talking not merely journalistic vindication for Jake Camden here, we're talking *financial security*. I guess that doesn't mean much to a guy like you, Max, who's had plenty of money all his life—but for me, it's a dream, an incredible opportunity."

"Yes. I see now. But how can I help?"

"Simple. I worked through the *Intruder* shysters from Jump

City last time I wrote a book and got totally screwed for my good faith. I need a good agent. Maybe a word in the ear of your publishers. Who knows. I hear you've got a production company now. Maybe you'd like to option the film rights. I need to get this ball rolling now, Max. Whaddya say? For old time's sake."

Schroeder looked down at his coffee. Seeing it was empty, he poured himself another cup. "It's decaf, you know. I had to give up the other stuff. The caffeine was wrecking my nerves."

"Max, don't leave me hanging like this. I spill my guts to you, and you dick me around?"

Schroeder smiled gently. "Jake, you're talking a great deal of blue sky. You have no manuscript . . . You don't even have an article finished."

"I've started it . . . It's going to run first in the *Intruder* as a series."

"Ooohh. Not good for your credibility."

"I swear, all the facts are going to be straight, Max. It's going to be a *great* series. It's going to be in the *Intruder* only because I need some quick money."

"As usual. I still don't believe you've paid back the several hundred dollars you've palmed off me at various conventions, have you?"

Camden got up and started to pace the room frantically, hands thrust into his pockets.

"I swear to God, you'll have the money back, and more! Max, please!"

Schroeder laughed. "Jake, of *course* I'll help you! Calm down. If all you want are good agents who sell UFO books, that's no trouble at all."

Relief flooded through Camden. He went to Schroeder and put a hand on his slight shoulder, patting him gratefully. "You won't be sorry, Max. I swear you won't. Look, I'll put you in the acknowledgments. Shit, maybe I'll dedicate the book to you. God knows, I have you in the text as the prominent authority on human-alien contact in the world."

Schroeder shook his head sadly. "This is not what I want, or what the Others particularly want, Jake. I'm afraid that you, in all your aggressive glory, are exactly what they fear so much in the human genotype. But the time of their coming draws near."

"Well, just tell them to wait until a couple years *after* my book and movie come out, so that I can be rich as well as peaceful!"

Schroeder sighed, as though in pity. "Jake, you've read my books, haven't you?"

"You bet. Stole a hell of a lot of ideas for stories from them, too, and thanks very much!"

"Yes, I suppose ideas *are* in the public domain. And I suppose that newspapers like yours do their share to disseminate important concepts. Still and all, when you listen to me speak at the conference, don't you think: My goodness, maybe I should think about who I am in the greater scope of the universe.

Do you think that the money I had truly satisfied me, Jake? Do you think that material things are what truly matters?"

"Hell *yes*!"

"Well then, take my word. They don't. I have received so much more from my involvement with them than I can fully express." Schroeder's blue-grey eyes turned away, growing reflective. He was in his early forties, and yet no trace of wrinkles had yet touched his face, no sign of skin hardening or drying marred his features. Yet somehow, he did not look young at all. In fact, thought Camden, he looked *old*. "I have received a peace, Jake. Like the Bible says, 'a peace that passeth understanding.' What will be, will be. I am only a vessel to work out the will of much greater powers than myself."

Camden got kind of spooked when Schroeder got like this. He'd seen it before, when the man would hold workshops at the UFO conferences: Merging with the Aliens. All sorts of spiritual mumbo jumbo, sure; but trembling at the center of it was a hypnotic *frisson* that made Camden decidedly uneasy. Camden liked the hard nuts-and-bolts part of the UFO business; chasing ghost lights, the imagination involved, the excitement of *just* maybe meeting with multi-eyed critters from light-years away. But all this spiritualist, channelling sort of bullshit kind of got to him. This showed in his stories; his "Shirley MacLaine" sort of puff-pieces were written with a decided tongue-in-cheek. And he was sure, in his columns on the subject, to make certain his audience knew his disdain. Nonetheless, it was just this "Elvis-is-Jesus" type of spirituality about some of the saucer business that people gobbled up. This was part of the reason that Maximillian Schroeder's books were so popular. They had a kind of exciting, pop, East-meets-West feeling to them, a Werner Erhard and Alan Watts take a trip on a flying saucer sense. People who got caught up in their mystic spell seemed to be able to look up at the stars and be connected in some way to the "cosmic mystery," as Schroeder called it. There were already small groups of "Otherness people" beginning to meet in churches and libraries to discuss the books, as well as their own experiences with the Others. Could a regular religion be far away? This was something that bothered Camden a lot about Schroeder, this evangelical, spooky aspect.

Still and all, he was a damned nice guy. And he was going to help him, right?

"So what are you saying, Max? You help me, and I gotta preach the will of the Others."

"No, no, of course not, Jake. I'm just pointing something out to you. Jake, sometimes I see in you a man hell-bent on his own destruction. I've seen you late at nights at the conventions, Jake. Drunk, your nose running from cocaine, a girl on your arm—maybe two girls. I wonder sometimes if too much money might light both those ends of your candle with a blowtorch."

"Hey! I can handle myself. Besides, I've given up the white stuff, and I can handle the liquor. Basically, I'm a reporter, Max. I want a story. This is a *great* story, which might just lead me to other great stories. That's my true habit pal. So cut me a break, huh?"

"Sorry, Jake. Perhaps I do preach a little too much. I apologize. I am merely concerned."

They chatted a little about mutual acquaintances, about the success of Schroeder's film, about his new novel he was working on, entitled *Redemption*.

"Well, speaking of that book, it is getting on, and I should get back to my word processor." Schroeder got up, got a piece of paper, wrote down some names and numbers, and handed the note to Camden. His literary agent, his publisher, his Hollywood agent.

"Thanks, Max. You're terrific," said Camden on his way out.

"Glad to help you out, Jake. Oh, one thing, though. You know, I don't want to exclude the possibility that Great Light—that's my production company—might not want first refusal on an option."

Jake grinned. "You old devil. Now you're talking my language. Sure, pal. You're scratching my back, I suppose I should scratch yours. I send you the articles as they come out, and tell the agent to give you first refusal."

"That's not quite what I mean, Jake. Things move very quickly in this business. This Scarborough business is ongoing. Much of it, I presume, is not in newspapers. I would like you to check in with me regularly, and keep me abreast of events. Would that be too difficult? You have my private number, I know . . . You arranged this meeting on it."

"You want me to tell you what's happening with Scarborough huh?"

"He is the hero of the unfolding tale, is he not?"

"Sure, Max. Sure, no problem."

"Would every other day be too difficult?"

"Gee, Max, sounds like you *are* optioning the story for your production company."

Schroeder thought about that for a moment, tongue poking out the side of his cheek. "Perhaps that would not be a bad idea. A small advance, put toward the welfare of a friend . . ."

"You option me now, Max, and I'll call you *every* day! I'll arrange a personal interview with Scarborough!"

"Yes, yes, you do need money, as you said. The interview, of course, won't be necessary. As a matter of fact, perhaps it would be best if Scarborough was not even aware of my involvement. We, after all, do not have a particularly pleasant relationship."

"No problem. Geez, and this will get me the agents right away. Max, you're the greatest."

"Maybe I'm just gullible."

"I promise you, pal, you *won't* be sorry."

"I sincerely hope not. And Jake—every other day will do just fine. I'll look forward to your call the day after tomorrow."

"You bet!"

They shook hands on it and Schroeder let him out.

Camden practically sailed down the sidewalk, avoiding the

piles of garbage awaiting collection. He'd gotten *far* more than he'd dreamed. He was right—this *was* a hot story. Otherwise, why would a guy like Schroeder want to snap up the movie rights. Shit, if he'd been a publisher, he probably would have bought the book.

Camden turned onto Lexington, headed for the subway, and his hotel. Mission more than accomplished! Yes indeedy!

He was halfway down the block, when he noticed the headline of the new edition of the *New York Post*.

"SUBWAY VICTIM IDENTIFIED! Fugitive Scarborough Implicated by FBI."

Shit, thought Camden. Shit and double shit.

He bought a paper.

CHAPTER TWELVE

The man sat on the cement bench, watching a fountain spurt in the middle of a concrete park, steeling himself for what he knew was ahead. Pigeons cooed and strutted along by an overflowing trash bin, pecking at moldy pieces of bread. There was the smell of sauerkraut in the air; a man sold hot dogs on the corner. Above, towered a colossus of tower and steel, dwarfing the human beings below, extraordinary in its own right, yet commonplace here in Manhattan, a tree in a forest.

The air tasted of the Hudson River, and of death.

This wouldn't be difficult, Everett Scarborough thought, trying not to nervously fidget in his shadowed place, face buried in a copy of the *New York Daily News*.

It was after noon. One of them would be coming out for lunch any time now. One the publishers of Quigley Books. He knew them all, and any of them would do. He would simply follow them, get them alone, stick this gun he had in his pocket in the small of their backs and suggest they take a walk.

Then he would get the truth.

Then he would find out where Diane was, and maybe something about *what the hell was going on*!

He watched the revolving doors carefully. Any moment . . . Any moment now . . . His hand was clammy on the cold metal of the gun in his pocket. It was the CIA man's gun, the one he'd shot. He'd sworn he'd use it only in self-defense, keeping it locked in his glove compartment.

Now, though, he just didn't care. He wanted to hurt somebody. Kill them. Get to the bottom of this. Find out about Quigley Publishers. Surely, it was a front for the CIA, the government . . . Maybe it was the headquarters of these Publishers and Editors that he'd heard so much about. He had to find out in whatever way he could, even if it meant violence. How long could he run? Not long, surely. They wouldn't expect this, no. They wouldn't expect him to take the tiger by the scruff of the neck and give it a good shake.

They wouldn't have expected him to barge into Donald Montcalm's office up there on the 33rd floor, either; they wouldn't have expected Everett Scarborough, wielding a deadly automatic to pull an Executioner-stunt like that. Or better: like the Immolator. Yes, Eric MacKenzie's men's action-adventure hero. Scarborough could almost visualize himself firebombing the offices. Assistants and art directors running for fire exits, hair aflame; editors and publishers crashing though plate-glass windows, flailing in the sky as they plummeted to their deaths. Ah yes, every wounded author's dream, that. ''Kill the fucking bastards!'' as the Immolator would say. ''Burn the villains!''

A trip up on an elevator would have been foolish. A word to the police or the grey suits, and that would be all she wrote—they'd have him all sewn up. But last night, in the heat of his anxiety, grief, and fear—last night, that was exactly what he was planning on doing. This modification—catching one of them outside—had come later.

He turned his eyes back down to read the Pete Hamill column again. It was close to the top of the newsprint, and he could keep that revolving door of the Broadway address in sight.

One of the assholes would be coming out soon. Maybe two—he could accommodate two. Large parties he'd already nixed. No, one or two would do just fine.

Last night, after the shooting, he'd lost himself in the Times Square crowd, escaping the shuddering glare of the rescue trucks, the police cars, the memory of those squealing subway brakes, the screams of the bystanders, that cold grip on his shoulder. He'd found himself in front of a 42nd Street movie theater, one of the sleazy sort that showed triple feature horror or Kung-Fu films. He'd paid the six bucks they wanted and sat through half of *The Laughing Dead*. The movie's zombies, decapitations, and general gory mayhem did little to help relieve his tension, but it was a good place to hide while the heat blew over. When he couldn't take anymore, he left and walked all the way over to Grand Central. There he caught a train downtown, where he found a cheap residential hotel near Chinatown called the Apex that had a halfshorted-out sign marked Vacancies on his smudged window. The sallow clerk took cash, asked for no names, and made no other kind of inquiry, which was fine by Scarborough.

His first reaction had been the desire to run; to get to his car, to leave Manhattan. Return to the comforting, anonymous road. The stay in that ammonia-laced theater gave him time to reconsider. No, he decided. He had to go to Quigley. The grey suits had tipped their hand. One of the strings that had tugged upon the Amazing Dancing Scarborough Puppet had been Cindy. Surely it had been her bosses pulling *her* strings.

The night had been restless. He'd gotten little sleep, tossing and turning in the threadbare sheets, the smell of mildew ripe in his nostrils. The man who'd warned him off, who seemed to be following him. Who was he? The program search looped in his brain, exhausting possibilities, keeping him awake. But then, almost at dawn, he'd fallen into a deep and dreamless sleep—a sleep that he hadn't awoken from until ten o'clock.

Come on, he thought, gazing up at the revolving door. Hurry up, you assholes.

A sudden thought struck him, making him squirm. What if they were all having one of their goddamned eternal editorial meetings. But no, probably not . . . Most likely, since the news as to Cindy's identity had hit the papers late, they were undoubtably pretty upset.

"Come on, folks," whispered Scarborough through gritted teeth. "Come out and get one of your three-martini sushi lunches!"

He had bought some scissors at an all-night drug store and lopped off the rest of his dyed curls. He hadn't shaved and a heavy growth of beard now coated his cheeks and chin and neck. He wore his glasses, so he didn't look like the pictures they had of him. He'd lost a lot of weight, too; and without his usual tailoring (and of course, with the other significant cosmetic changes), he figured he was safe enough out in the open. He didn't think anyone would recognize him and turn him in.

"Scarborough. Hey, Scarborough!"

Never before had the sound of his own name alarmed him so much. And close . . . so close. He spun about, prepared to either launch himself at an attacker, or sprint away.

When he turned, he saw the abrupt sight of a slender man in a jacket and a Hawaiian shirt coming his way. The man slid onto the bench, sidling up next to him. "Thank God I caught you here. I figured there was a good chance you'd come here."

Scarborough blinked, aghast, and yet relieved at the same time. "Camden! What are *you* doing in New York?"

"I could ask you the exact same thing, pal. But me? I'm on assignment from the *Intruder*."

"Go. Leave. I've got something to do here. Something very important." He tried not to sound desperate, but he was all too aware of the way his voice husked.

Camden gave him a disappointed look. "Evvie, Evvie. This isn't too smart, chum." He looked around, chewing on some gum, gazing out at the people walking up Broadway. "Someone could see you here, out in the open. I'm just glad I caught you before you went into Quigley. I read the article in today's paper, man. Pretty sad stuff. The editor of yours . . . she was one of *them*, huh?"

"Yes."

"Well, all this makes for a juicier story, I suppose, but you're going to do neither me nor yourself any good by hanging around Quigley Publishing."

"One of them will tell me . . . Who knows. Montcalm. Tarrants. *They* can tell me where Diane is . . . They can tell me who these Editor and Publishers are."

"Don't bet on it, chum. Look, I feel a bit . . . exposed out here. I know a nice dark place a few blocks away where we can have this discussion in private. What do you say?"

"Just go, Camden. This is something I've got to do."

"Look, Scarborough. I'd hoped you'd be cool, but you clearly aren't. Chill *out*. Okay, so this Cindy Editrix was *One of them*. In that case, she was planted. No way Quigley Publishing could be owned by the government or any shady subsidiary. Think, pal. It's part of a gigantic corporation. Some huge German company, right? These guys could *buy* the government!"

"Maybe they did." But Camden was right.

"Uh-uh. These folks are out to make *money*. It's called 'business.' Now, I realize we're getting a tad paranoid, but don't you think we should at least have a little lunch before you start shooting up Broadway?"

Scarborough thought about this a moment. He wasn't so sure, but Camden had actually introduced a thought that had to be counted as a logical possibility. Quigley *could* be innocent of these particular operations, using him. Cindy and the Editors/Publishers could have simply *used* Quigley just as they used him.

"But if I can just talk to one of them for a moment," said Scarborough, feeling the shaking doubt in his own voice beneath the frayed determination.

"Please! Look, I'm begging you. At least give it some thought? You're wound so tight you look like you're about to break. C'mon, catch these guys coming *back* from lunch."

Scarborough's pride hated to admit it, but Camden was right. There *was* the possibility that his brain was a touch addled from all the drama last night, his lack of sleep, his feelings of desperation concerning Diane, concerning his whole life. Give it a couple hours, he thought. Give yourself a chance to get a grip.

"Okay," he said, taking his hand out of his jacket pocket.

"Good. So what have you got in there?" Camden craned to check, and caught the form of the gun. He gave a David Letterman grimace. "Yo, Dirty Harry. Don't you think that your heat there's a little bulky. It's broadcasting at about 50,000 watts!''

Scarborough looked down. Camden was right again.

Christ, what could he have been thinking about? What was wrong with him? What was going on inside of his formerly precise and ordered brain?

Quietly, he took off the jacket, tucked it into a discreet ball in such a way as to hide the unsightly outline of the automatic weapon.

''All right!'' said Camden, looking around as though catching a bit of Scarborough's paranoia. ''We're outta here!''

If New York City is surely a place to drive its residents and visitors to drink, then it obligingly provides them with numerous watering facilities. Frank Sinatra sings of waking up in a city that never sleeps. He perhaps should also have mentioned the difficulties of drying out in a city whose bars never close. Licensing hours in Manhattan run from 6:30 in the morning until 4:00 the following morning; and if you're still feeling sociable at last-call, you might wobble over to one of the illegal drinking establishments, often as not run by Irishmen.

Most visitors to the Big Apple—and indeed, most of its well-to-do inhabitants—drink at the pricely, trendy bars such as Elaine's and the 21 Club. But if you want to get the real flavor of the area and haven't got four dollars for a martini or five for good spirits, there are always the Irish working-class bars. Camden had found them on his very first trip to New York. They were places like the Blarney Stone or the Blarney Rock or McAnn's—plain, big places that have probably sold more drinks over the years than McDonald's has hamburgers.

Camden and Scarborough sat in one of these Irish bars now, huddled together in a booth in the far reaches of the dining room. They'd found the dim space just before the lunch crowd started coming in, and Scarborough had to admit that it was a good idea coming here. The likelihood of being noticed in a place like this was close to nil. And besides, a big red exit sign shone just in a hallway just a few feet distant. A way to get out, quick, if necessary.

The Harp was typical of the Irish workingman's bar. It consisted of a creaky wooden room with a long, ancient bar running down one side, stocked with a plethora of bottles capped with bright metal spouts. Shot prices were bannered from small signs in magic-marker script. For instance, the special today was a shot of Old Crow for just a dollar and ten cents. Other shots, however, were not a whole lot more, and as Camden had rapidly learned, at these sort of places the shots were generous.

Besides drinks there was the steam table, where you could get a good corned beef sandwich with a side of cabbage, or a roast beef platter with potatoes and two vegetables for not much more. Wisps of steam rose up past the rosy-cheeked face of the middle-aged man cutting sandwiches now, as he probably had been for years and years. The place smelled of spilled beer and old food, but it was an honest, homey smell, like grandad's boots.

When Camden brought back his shot of Johnnie Walker Red and a big frothy pint of draught Guinness from the bar, Scarborough had hardly touched the mug of New Amsterdam Amber that Camden had brought to him first.

"Hey—have a couple of sips at least. It'll take that metal rod out of your spine!"

Scarborough sipped tentatively. The beer was cold and good, with a hoppy, grassy taste similar to the stuff you get from the microbreweries, only not so pronounced.

"Good stuff," said Camden. "Made right here in the city. They even have brewery tours. Drink up. Hell of a lot better than a Valium. Of course, if I'd had one of those, I would have jammed it down your throat back there. Cripes, you want to get yourself nabbed—or worse, *killed*?"

Scarborough turned and looked around nervously. They were well back, though, from the rest of the half-full crowd, and the babble of conversation overlapped the bump and flow of jukebox music. Satisfied that they weren't being overheard, he turned back to Camden. "You don't know what I've been through."

"I can guess. So drink up, I'm telling you. The number one cure for tension is alcohol!"

"Haven't had any of this stuff since Vegas. No—before. Maybe I'm afraid if I start, I'll never stop."

Camden sipped his beer, letting his shot sit. "A couple beers won't hurt. Look at me now. I'm not going to drink this whiskey here until I have something on my stomach. So what gives?"

Scarborough took another sip. Camden was right. This was relaxing him. "Jesus, I'm a mess," he said. "Thanks for heading me off at the pass, Jake. You're probably right." He relayed in detail his meeting with his editor, and his resultant discordant feelings and distrust of people. "Who would have thought, Jake, that you're one of the few people I trust now."

Camden whistled lowly. "Rough stuff. I'm sorry. It'll all come out for the best—we'll find Diane. Meantime, catch me up. I take it you drove all the way out here."

"Yeah. Parked on the West Side of Manhattan. I've been trying to call you, Jake."

"Well, you've got me now."

Scarborough outlined the trip across the country.

"Fine," Camden agreed. "I'll check in with Marsha. She'll keep us in contact from now on. I'll be a faithful correspondent. What say we get something to eat now."

Jake got a meat-loaf sandwich. Everett took the corned beef on rye, with a side of chips. Camden had been right. It *did* make him feel better. He had another beer as well, and it did the trick. He actually was starting to feel better again.

"The thing that's got me flummoxed is the guy who hurried you out of the subway," said Camden, after putting away his shot. "You sure you weren't hallucinating?"

"No. And I'd noticed him, before. He was with another guy before. These two guys are following me, Jake. Remember how I couldn't figure how I'd nailed that CIA guy—well, I think maybe I didn't. I think one of my guardian angels did."

"It doesn't figure. They sound like agency people. They sound like pros all right, though. What do you think? Warring factions? Maybe guys from the NSA, you think? Foreign agents? Cripes, could be several possibilities and none of them pan out."

Nonetheless, Scarborough noted that Camden had a slight smile on his face. This clearly was getting to be a better and better story, and the bastard was actually *enjoying* himself at Scarborough's expense. No, no, thought Scarborough. Mustn't get myself worked up and paranoid. Whatever else may be going on, Jake Camden had proved to be a valuable and trustworthy ally.

"That's why I need this information. I mean, what do we have? Some kind of conspiracy, sure—but *why*? What are they hiding—and who are they, exactly. I mean, we know they have government links—but there's more. I feel it in my gut."

"Dr. Everett Scarborough, getting hunches. Is Mr. Logical perhaps becoming clairvoyant."

"Don't laugh. Some strange things have been happening."

"Okay, let's put the facts together . . . I mean, the ones we have, anyway." Camden ticked them off one by one with his fingers. "Project Blue Book was a cover-up. Your daughter and her boyfriend were kidnapped by whoever perpetrated that cover-up. The cover-up appears to include a good deal of false UFO sightings strewn among the genuine—as well as falsified info concerning document evidence."

"Disinformation."

"Yeah. Anyway, not only were you a prime scientific patsy for those years you worked on Project Blue Book, but they stood you up as an antiUFO skeptic, helping you forge a career that brought fame and fortune. Again, a variation on a theme. Question is, why didn't they just buy you off?"

Scarborough glared at him. "Maybe *some* people *can't* be bought off."

"Okay, okay. Maybe you need another beer. We're just trying to look at the facts and speculating." Jake drank another finger's worth of liquor. "They're so worried about MacKenzie's full Blue Book files and other research, they torch it. And kill him when he barges in. Nice guy, too. I liked him."

"Yes." Scarborough looked down at his mug, touched by grief again. "He was the best." Maybe he *did* need another beer.

"Okay, they know you're onto them—and thank God they don't seemed to have glommed onto *my* activities—and so they try and kill you. Doesn't work. So off we go on a merry chase. Fugitive time, only you're chasing ghosts, not a one-armed man."

"There's something you've left out."

"Hey, hey. There's my man. You're not going bugfuck after all. You care to iterate said 'something.' "

"That farm. Those abandoned farm buildings looked as though they used to house some kind of scientific testing sight or something."

"Dead on, Evvie. *That's* the stick they got stuck up their beehind. *Everything* started happening after that. That caretaker—the more I think about him, the more I think, used-up spook. They stuck him there to guard the place. What else could it be."

"Yes, but what could they have been using it for. I've been racking my brains, trying to come up with that answer. I even thought about stopping back in Iowa . . ."

"Wrong. You don't blow up the same bull's ass twice."

"I didn't, of course. But I thought about it. Are they hiding some kind of special scientific project, I thought. But what could that have to do with this campaign of disinformation? It just doesn't make sense. There are too many pieces missing."

"You didn't have much to do with this Colonel Dolan fellow and what's-his-name, the guy Marsha mentioned to you who was yakking about the Publishers and Editors. Oh yeah, Richards. Brian Richards."

"I did not *know* the man. My sole contact with the military establishment was Colonel Dolan. I worked with some people in other departments for research, but that's all that was—research. Camden—I was just a writer . . . a citizen . . . My days *inside* the military—such as they were, ended twenty years ago. I simply wrote and talked and played the media personality—and that's all they wanted from me. God, what a *fool* I was. Camden, the more I think about Cindy, the more I think about the whole thing, the more *violated* I feel, the more incensed, that these bastards have irreconcilably pissed upon my rights as a citizen—to say nothing of everything that this country stands for."

"Quick. Drink the rest of this beer. We really can't afford a scene here, Scarborough."

Scarborough took in a deep breath, exhaled. "Sorry."

"So, what it all comes down to is *what* were these people doing in those labs on the farm? Apparently, it's been going on for a while. For one thing, it seems they were faking UFO reports. Disinformation. Both you and I know the standard, saucerite-paranoid rap on that line."

Scarborough sighed. "Yes. Disinformation to cover up actual contact with saucers. By the government. Lunacy! Did you read that article in *UFO Universe*, the one that came out just before all this started? The one with that fellow claiming the government made some kind of *deal* with aliens, trading rocket secrets for rights to experiment with U.S. citizens? Even with all that's happened, I'm still very far from buying that line. No, what we have here is your run-of-the-mill government cover-up. We're not getting into *aliens* yet."

"Not even those guys who are following you."

"They look normal enough to me. Maybe they're members of some other counterintelligence agency. . ."

"Look, let's get back to that lab. Seems to me that you've been through far too much to put down paranoid ideas. Stretch your thoughts a little—"

"They've been stretched. You think that these people aren't just faking alien saucers. They've been faking alien abductions. Budd Hopkins stuff—Strieber—Schroeder."

Camden cleared his throat, looking a little uncomfortable. "Yes. That just what I mean."

"Okay. But *why*?"

"*That's* what we've got to find out, buddy. Whatever it is, it's *outrageous* from the word 'go'! No wonder they've been bending over backwards so suddenly to discredit you. They know that if somebody legit comes out with the facts, the idea that a government-associated group—I don't care if they're the CIA or the White House janitors—are *kidnapping* people, fooling with their minds . . . Well, that's going to create a *colossal* stink, to say the very least. Who knows what high mucky-muck heads will fall. Sheesh, maybe the whole government!"

"Let's not take it too far, Camden."

"I'm serious. And Scarborough, if we can find out what's going on—get the proof . . . Hell, just the *alleged* truth is enough—this will be your salvation and probably Diane and Tim's as well."

"I'm not quite sure I follow."

"The Power of the Press, man! You will be vindicated, and these guys won't be able to *touch* you. Maybe that's what these guys who are following you are working on . . . Yeah, maybe they're discontented spooks, trying to foul up these cancerous goings-on, trying to do the right thing without giving their identities. What do you think?"

Scarborough grunted. "I like that a whole lot better than the aliens stuff."

"Okay. So that leaves us a good deal ahead of where we were before. Certainly better than having you accost a poor bookish executive with an automatic weapon!" Camden finished his beer and stared back at the bar thoughtfully. "You know, all this talking is really making the old brain work."

"Something you've left out?"

"No, something I remember seeing. I'm a member of a clipping service, Scarborough. You know what they are?"

"Of course. They save articles from all publications. Local as well as national. International, if you want them."

"Exactly. Well, this service clips out news items concerning UFOs. Invaluable in my trade."

"I should imagine. Yes, I've heard of that service."

"Sorry. Keep forgetting we've been working the same fields for a while. This is a service, though, which I'm sure you didn't subscribe to. It sends the stories *immediately*. Quite necessary for a timely publication like the *Intruder*."

"Sounds quite expensive as well."

"Yes, but well worth it. Anyway, there's this one particular story I was reading on the plane here last night. Just came in. I think you should have a look at it. It's back at my hotel room."

"I think maybe you should call Marsha and read it to her. I'll contact her this evening."

"Not a bad idea. I assume you're leaving town then."

"Yes. I probably should have left last night."

Camden chewed his lip a moment, as though trying to figure out the right way of saying something. "Look, man. I know this is all pretty tough on you, and you have to take the brunt of all this on your own shoulders. God, it must be rough to have your whole world torn apart like this. Jeez, your daughter kidnapped . . . I guess we ain't had exactly the warmest of professional relationships and all, us being from different sides of the camp and all. But in this situation, you just gotta know that I'm in there, rooting for you. And I'll help you whatever way I can."

"Thanks, Camden. I guess you did me a favor back there. I was really out of it. Maybe the first thing I should do is check into some hotel in the sticks and sleep awhile."

"Who knows. Maybe your two guardian angels would have come around and kicked your butt. By the way, where are you going?"

"I have to head back to the D.C. area."

"Isn't that a little dangerous?"

"Not any more than New York City."

"What's in D.C.?"

"I've got a friend who works in the CIA. A field agent. He's on record as an individual not particularly in favor of illicit activities of the Company. And he owes me a favor. I've been trying to reach him. Maybe I have to go down and visit him personally."

"This guy have a name?"

"Why do you need to know?"

"Yow! Get thee behind me, Paranoia!"

A weary smile crossed Scarborough's face. "Yes, maybe you should know, Jake." He finished the last dregs of his beer and clopped the mug down onto the grainy table. "His name is Edward Ross Myers."

CHAPTER THIRTEEN

"Pardon me, Lieutenant Manning? I don't mean to bother you, I just have a new file of documents for you, pertaining to your present assignment."

The knock on the door had already startled Marsha Manning, but she didn't have time to put her book away before Captain Fredricks marched in, a manila folder tucked under one arm, a tentative smile on his face.

"Captain!" said Manning, turning the large flat hardback down onto a pile of printouts. "I was just taking a coffee break! Fascinating stuff, this statistical work!"

"Thank God for people who know computers, that's all I can say," said Captain Fredricks, the twinkle in his eye matching the gleam that the strip lighting made on his bald head. "Without you folk, I guess we'd have to march the troops out and invent some kind of drill involving fingers and toes, eh?"

Manning laughed hollowly. "I must say, I do prefer field work."

"So I hear. . ." The captain bent his neck, fingered the spine of the book. It was a *Time-Life* book called *The UFO Phenomenon*. "Reading up on your aircraft, eh?"

"I realized there were large areas of ignorance concerning the subject, sir."

"All bullshit of course." He gave her an appraising look, raising his bushy eyebrows. He was getting into his mid-forties and those eyebrows were turning grey, which gave him a rakish appearance he liked to use on women. "I've piloted for years and I've never seen a one. Mind you, I've seen a few odd things in the skies. Sun dogs. Moon dogs. Ice crystals. All explainable. Atmosphere does funny things. So does lack of pressurization, lack of oxygen." He handed her the folder. "Well, then, you think you can incorporate this material into your sublimely athletic algorithms?"

Manning took the folder. She flipped through the thick sheath of xeroxed reports on the mechanical functioning of various jet and propeller engines utilized by the United States Air Force over the last thirty years. This kind of stuff was the sort of thing the Pentagon usually farmed out to contractors in Washington D.C. Giving her this assignment was the Air Force's form of retribution. They couldn't pin anything on her, but they had their own sort of punishment, anyway. Rather like Chinese water torture, except the "drip drip drip" here was "byte byte byte."

"Thank you, sir. I believe this will help on the attrition tables."

Captain Fredricks cleared his throat officiously and then turned to leave. He stopped himself, though, and turned back to Lieutenant Manning. "I suppose you've heard the stories about this air base."

"What?" Manning blinked, playing dumb. "A UFO landed at Wright-Patterson?"

"No, no. I mean, about the crashed saucer. The one that was supposed to have gone down in New Mexico in '47. Near Roswell, I think. Well, the myth goes that we're keeping it here in some hangar, along with the recovered bodies of dead spacemen."

"Sounds a little wild, doesn't it, Captain. Is it true?"

"Of course not. I just thought you'd be interested in a little local mythology."

"I guess the place does need every little bit to liven it up, doesn't it?"

"That it does, Lieutenant. That it does. When do you think we'll see the preliminary charts on these statistics?"

"What, the graphs? Well, I've got two weeks, but unless there's more stuff you want me to work with, and unless I find I have to recode the program to suit this material, I suppose I'll have it finished by. . . about next Tuesday?"

"Oh, no hurry. No hurry at all." The captain smiled a faintly perturbed smile, bid farewell, and left. Clearly, the man had only received hints of what she'd been up to a couple of weeks ago and was curious—was fishing for juicy details. All he knew was that suddenly, she'd been figuratively chained to an outlet of a mainframe, doing very boring detail work, far, far beneath her abilities.

But then, this *was* the military, wasn't it?

Marsha Manning was a career officer, and generally speaking, when you'd chosen that path for your professional career, you either loved being part of Uncle Sam's Defense Family—or you were planning a new career after you served your twenty-five years. Manning was in the latter category. She wasn't looking forward to actually being in her mid-forties, but she *was* looking forward to her new career. In the meantime, she'd been trying to figure out a way to actually experience something akin to being alive. Meeting Everett Scarborough, getting involved with all this, was an exciting roller coaster, but now she was back in mundanity. She wanted to get on that 'coaster again. She liked the feeling, and she liked Scarborough.

Manning tossed the packet of papers contemptuously into her in-box, and then stared back at her computer screen. She'd called a subroutine of the program up for a little fine-tuning. It was in *C* language and she diddled a little here, a little there, and then quit. Hell with it. It certainly wasn't the most elegant subroutine. But then, elegance was lost on the basic program anyway, which was clunky as hell. Your really *good* programmers and analysts were out in private industry, pulling down hefty chunks of change. Air Force programs tended to wear hobnailed boots and crash a lot, like Air Force pilots lately.

"Shit," she said disgustedly, picking up her cup of cold Maxwell House and sipping bitter reality, without cream. She stared out the dun brown window onto a bleak day across the plain of architectural nightmare which was the Wright-Patterson office complex. A Boeing jet was taking off one of the runways, screaming into the sky to tear hell out of the sound barrier and burn a lot of fuel, probably for no particularly good reason. A light rain had fallen early this morning, and clouds still hung low over the base. The grey blandness of the cement and the buildings seemed to seep in around the window sill, smelling of tarmac and oil and khaki pants. The smell that Manning had lived with for a long time, which now weighed down upon her, like those clouds up there pressed down upon dull Ohio earth.

Marsha shuddered with an imaginary chill, feeling oppressed, feeling claustrophobic.

"Shit," she repeated. Then she went and closed the door of her small office cubicle, locking it so no more captains would barge in with barely a knock.

She picked up the *Time-Life* book again, and turned to the index. Sure enough, there it was: "Wright-Patterson Air Force Base." She turned to the indicated page and then read the passage. She read the page before the paragraph mentioning the base, and then the page after it.

Hmm.

Apochryphal information, of course. Hearsay, rumor . . . myth, just as Captain Fredricks had said. Nonetheless . . .

According to the story, an Unidentified Flying Object had crash-landed in some rancher named Brazel's land in southern New Mexico, close enough to an air field that military personnel were able to go check it out. Here was where the story split off into a collection of odd newspaper clippings. Initially, it was said, the commanding officer was supposed to have let it be known that it was a flying saucer that had landed, and the military team had also discovered the bodies of "aliens." There were reports of wreckage consisting of a strange and tremendously strong foil-like metal. Suddenly, however, the official Air Force line changed. No saucer, no aliens, they said. It was just a weather balloon, newly tumbled from the heavens. That's it. Really. Honest! Actually, from what Manning read, that was indeed probably what it was. She could see how easily such a story could snowball into rumors afterward. How this "saucer" and its contents made it to Wright-Patterson, the authors did not mention.

Still, it wasn't like she wanted to go back to working on that stupid statistical program—next thing she was supposed to do was to start inputting the information, and that was close to just secretarial work. She dreaded the thought. Besides, it had already been proven to her that the Air Force could be duplicitous; Scarborough's dealings with them showed that plain and bold. Maybe there was something to this story, some scrap of truth that the boys in brass *were* covering up.

Wouldn't hurt to look, would it?

Besides, it was a challenge.

Smiling, Marsha spun back to her computer keyboard. With the password (which she hoped they hadn't changed yet) she accessed the mainframe.

Phosphor dots wriggled, letters and numbers jiggled.

Suddenly, she was through. She was on-line.

As she recalled, there was some sort of menu in this system where'd she seen a storage section. All she had to do was to wade through the byzantine corridors of this clumsy system.

If something was there, she'd find it.

Even if she had to work through her lunch hour.

The sky, if anything, had gotten darker. An occasional fat droplet fell down, splashed onto the ground or into one of the many puddles scattered about like sad little mirrors reflecting the grimness above. Marsha avoided these puddles as she walked the network of sidewalks and parking lots, past the grids of buildings toward Storage Area 34. She wore a standard-issue Air Force parka against a possible downpour. It wasn't serious yet, so she kept the hood down. It was drowned-worm weather, and it smelled of damp soil and grease on water. The squeal and roar of jets had dimmed somewhat—she was closer into the building complex, so the sounds were muted somewhat. Her flats squished slightly as she walked. A northeastern breeze flipped along the luxuriant growth of spring green grass where it had been allowed to grow, filling a tattered wind sock near a barracks that someone had put up as a joke—wind socks were hardly the stuff of your high-tech Tom Clancy Air Force, no sir!

In her pocket was a piece of paper upon which she had written a compartment number. Stall A–97, Room 18, Storage Area 34. She'd picked through the storage directory for close to an hour before she found the listing.

All it said was, "Roswell, '47."

Not "Artifacts," not "Flying saucer," and certainly not "Dead alien bodies." Nothing else. Just "Roswell '47."

Still, the fact that there was a listing like that was worth taking a look. Nothing else to do. Who knew what she might turn up that might help Scarborough, might help explain exactly what the government had been covering up all these years.

A map of the base had provided the location of Storage Area 34. It wasn't too far away, not driving distance anyway, on an older part of Wright-Patterson. Now, as she approached the area, Marsha could see that the buildings were old hangars and support-buildings. They looked particularly dilapidated and grey—the Air Force hardly built things to last a long time, and doubtless bureaucratic foul-ups had allowed these things to remain standing.

Nonetheless, as she walked along an old runway approaching them, as she located Storage Area 34, she was surprised to see that it was one of the two hangars which were surrounded by a fence.

An electrified fence, with barbed wire at the top.

There were no guards near the gate to the fence which wrapped around Storage Areas 34 and 36—but the gate wasn't exactly hanging open either, Marsha saw. In fact, there was a very modern-looking booth in which a computerized access lock was stationed.

Marsha stepped in to examine this, already suspecting what it was, but checking nonetheless. Sure enough, it was an authorization lock. You had to slip in your identification card and then put in the special authorization code particular to this machine. Just like an ATM bank-teller.

Damn. There'd been no mention of an authorization code in the computer.

And if she stuck her magnetic-strip in there, who knew! Her attempted accessing might just hit an alarm bell in security, and her name might zap out onto some printer, underlined and bracketed with asterisks. After the questionable actions she'd pulled with Scarborough, Colonel Walter Dolan had wanted her ass in a sling—only intercession from her higher friends in the force prevented a serious inquiry, the first step to a possible displinary action or even court-martial. Her buddies had suggested that she be a "good girl" for a while. Getting caught accessing classified material was not exactly going to give her that kind of reputation.

No guards, though. That was significant. Not even electronic surveillance. This particular classified area was apparently not all *that* important.

Not exactly "Mission: Impossible" territory.

She strode past the lock box, not even stopping to inspect it. She'd seen its type in other places on the base, knew its operation.

Maybe she should have expected this impasse. Whatever the case, the blocking box not only intrigued her, it *challenged* her.

Clearly, she was just going to have to go back, snoop a little through the security systems and do a bit of finagling with the codes.

With a private little smile, Lieutenant Marsha Manning swept past Storage Building Number 34 and walked on around the quadrangle, heading back to her office and the modest bag lunch she'd prepared for herself.

Yes, this was going to interest Dr. Everett Scarborough.

A lot.

CHAPTER FOURTEEN

The main headquarters of the Central Intelligence Agency is an imposing fortress of a building, modern in architecture, stark against a wooded area in a place called Langley, Virginia, across the Potomac from the rest of the government and a few miles north of the Pentagon in Arlington. Although the rest of the CIA Washington employees are scattered all over Washington, D.C., and Northern Virginia, with thousands of consultants in the science, technology, economic, government and defense community sprouted about the swampy area like mushrooms from dark, dank soil—to say nothing of the ten other agencies such as Defense Intelligence, the National Security Agency, and the FBI which report directly to the CIA—it is in Langley that the Director of Central Intelligence and the head officers have their offices.

Washington, D.C., is a hot and humid place. Before the advent of air-conditioning, the U.S.A.'s capital was considered "subtropical" service by other country's diplomats. Therefore, when the huge building was finished in 1961, it came as a surprise to the over 10,000 employees installed there that although the heating system was adequate, the air-conditioning system was wretched. Thermostats were placed in offices, and regulatory orders were placed on room temperatures. However, when continual manipulation of the things fouled up the AC even more, lockboxes were placed upon the thermostats. The contractors had not, however, reckoned with the fact that many of the CIA's employees had taken a "locks and picks" course during their training. Practically all the thermostats were back in operation in very short order.

The area housing Brian Richards's office had never had a problem with faulty air-conditioning. This portion of the CIA was where the generals of the agency were placed—the so-called super-grades. Civilian brass of the first water. Indeed, the super-grades were accorded a remarkable number of civilities and comforts, including their own private dining room.

Brian Richards sat in that private dining room now, having lunch.

Although the chefs here were quite good with French cuisine, the latest trend in gastronomic preference in Washington, D.C., was Italian dishes, and so the CIA upper-crust exclusive commissary adjusted accordingly. Now, Richards was enjoying, after his antipasto, a variation of *ribollita* called *pappa al pomodoro*. For dessert today, he intended to order the *zabaglione*, a delicious Italian egg-custard he particularly enjoyed.

However, he never got to his dessert that day.

Richards enjoyed the private dining room, which had gourmet food at modest prices. There were two other places to eat here: the small and dismal cafeteria for employees of other government agencies and guests, and the restricted better cafeteria which served the other CIA employees. This was so that visitors would not be able to see clandestine CIA agents having their lunch.

Richards had always detested that cafeteria, and the day he'd been promoted to a subdirectorate, he'd celebrated by ordering a particularly expensive Chateau Rothschild wine in this tastefully decorated place, which looked more like a fancy downtown restaurant than a commissary; then he'd drunk it all himself.

He was drinking no wine today. In fact, he was eating lunch faster than usual, and would have ordered just sandwiches in his offices if his pride did not disallow it. No silly ass like Everett Scarborough was going to *totally* disrupt a cherished perquisite of his office, no indeed.

Ivan, one of the black waiters in starched white shirts and ties, came up to him with a message. "It's your office, sir," said Ivan. "You're wanted back immediately."

"Did they say why?" said Richards, looking up with an annoyed expression.

"No sir. But they did say it was a priority."

Having a waiter at the super-grade commissary named Ivan was a constant source of jokes for the officers. The rumor was that George Bush had hired him before leaving, just to annoy the next director who came to power.

"Thanks, Ivan. My compliments to the chef. I guess I can't finish." He patted his mouth with the newly laundered, embroidered linen napkin and then stared down at the unfinished food on the fine china. "Excellent, though. Please put the bill on my tab, and give yourself an extra five percent tip."

"Thank you, sir. This *must* be important."

Richards smiled deprecatingly. "And guess what! It's top secret. Classified stuff of the very *highest* order."

Ivan chuckled at the joke. "In that case, sir, you'd best not say another word to a man named Ivan."

Richards laughed congenially and patted Ivan on his back. "That's right. See you tomorrow. Hopefully."

When he reached his office, his secretary—a man named Harold Finch, looked up from his typewriter and pulled off his dictation earphones. "It's Operative Edward Myers, sir. You told me to send for you as soon as he came in."

"Myers, huh? Well, well." Richards cleared his throat. "Harold, why don't you take lunch now."

"I've already had lunch, sir."

"A break then. A hard-working man like you always needs a break."

Finch nodded. "I understand, sir."

Harold Finch was paid not only by the CIA, but also from a private account by Brian Richards himself. He was paid for what he knew, and for what he did not know. Richards had no reason to distrust his secretary, but this meeting was something he should not even have a chance of overhearing.

Finch got up, and was calling the phone operator to make arrangements for reception of any incoming calls before he left. Richards cleared his throat, straightened his jacket, and then opened the door of his private library/conference room. Ed Myers was seated at the table, reading a magazine.

It was the gag around the agency, among those who came into contact with him, that Ed Myers was in the wrong profession; he should have started working for Walt Disney. He had open, blonde, California good-looks about him, a casual and lanky jokiness about him that creative types in Hollywood tended toward; smooth and slick and yet as all-American as a grown-up graduate of the Mickey Mouse club. He was tall and slender with short hair, cut at least once every two weeks, and when he shook your hand, as he shook Brian Richards's hand now, you got the feeling of, Hey, the world is an up kind of place, along with general twinkly-eyed enthusiasm.

Nonetheless, Myers could be a hard and competent operative, Richards knew.

"Well, thanks for coming around, Ed. Sorry to make it sound so hush-hush and urgent, but it is!"

"That's okay. Just got off the plane at National and had to head out here anyway to make sure my report got in the right pigeonhole." The man did look tired. Richards noticed that the wrinkles around his eyes were more pronounced.

"You were in Berlin, I hear."

"That's right."

"What, smuggling ballet dancers across the Berlin Wall."

"You know I can't talk about that kind of thing, Mr. Richards. Even to you."

"Good man. Of course. Well, could we adjourn to my office? I think I might be able to rustle up some coffee for both of us."

Myers brightened. "Do I look that bad? Sure, I guess I could use some caffeine. Not as young as I used to be."

"Who is, Ed. Who is?"

The two men walked back to the office. They both wore their laminated security badges of photograph and ID number clipped to the lapels of their coats. All just a part of the extensive security precautions taken at Langley, which included a barbed-wire fence around the perimeter of the building patrolled by armed guards and police dogs. No one could enter the perimeter without a proper ID-check and clearance. Even the janitors needed extensive security checks. This area was particularly cordoned off, with another check necessary to get past guards posted in glass booths in the hallway up ahead. There were those who claimed that the *C* in CIA stood for *clandestine*—but one acerbic wit had suggested that President Harry Truman, who had formed the agency back in 1947 with the National Security Act, simply couldn't spell *security*.

Richards gestured to one of the cushioned seats fronting his desk, and then shut and locked the door. The room was soundproof now, so he felt a little better. What was going to go on here today was not the sort of stuff he wanted *anyone* in this area to even have the vaguest hint of.

"How's the family doing, Myers?"

"Just great, sir. My son, especially. He's really come around."

"Good, good. Glad to hear it." Richards walked around his desk, then settled into his huge and quite comfortable leather swivel chair. The placement of his chair afforded him a pleasant panoramic view of the array of degrees, awards, and citations which papered his walls, his collection of books, his small locked filed cabinet, the pictures of his own wife and family, and the plush couch against the wall. A turn would give a beautiful view of the spring-clothed Virginia woods in back of Langley, but for now, his curtains were closed. The office was lit by tastefully placed electric lights; there were no harsh strip fluorescent lights above. Richards abhorred them.

"Families are wonderful, aren't they?" Richards continued. "What would we do without them. But now, to business." Richards brought out a small portable tape recorder and placed it on the desk between them. "I understand that you're a friend of Dr. Everett Scarborough."

"That's right. He's helped me . . . and my son. Quite a bit." A cloud darkened the man's features. "This isn't about Everett, is it?"

"It is." Richards cleared his throat. "I suppose you haven't been following the national news in Berlin. I've taken the liberty of having a few copies of news stories prepared for you to look at, Ed. But let me go over a few succinct facts in regard to the case."

Richards outlined the party line on the situation, summing up with, "He's apparently had some kind of nervous, perhaps even mental, breakdown. We don't know—that attempted assassination over at the University of Maryland may have triggered it. All we know for sure is that he's killed Captain Eric MacKenzie, and a top operator. He has interfered with top National Security Projects, and is running amok on some wild paranoid mission against the government of this country, claiming that we've kidnapped his daughter." Richards pulled a paper from a stack, opened a page and showed it to Myers. "The latest is this. Scarborough's editor at Quigley—Cindy Clinton—was killed last night. Shot and thrown in front of a New York subway train. We have reason to believe that Everett Scarborough was involved."

Myers's face had remained impassive through the entire briefing. Even his normally expressive eyes had remained unemotional, if faded and tired. Now, though, he shook his head. "Everett? You know, this is awfully hard to believe. Everett *Scarborough*? I suppose he can be a bit of a curmudgeon—but my evaluation of him was always that he was an emotionally healthy individual."

"Ours as well. As you probably know, Scarborough worked for the government in the sixties as a top consultant on Project Blue Book. After Blue Book concluded in 1969, Scarborough remained on-call—and also utilized Air Force and government facilities for continued information for his series of books and lectures that made him somewhat of a media star. His contact for these purposes was Colonel Walter Dolan. Our intensive interviews with Dolan have led us to believe that Scarborough has been headed for a crack-up for some time. Indeed, some grandfather memos from Colonel Dolan have been unearthed, expressing concern. It was a shame that we did not pay attention to this disintegration of a brilliant mind—but then, there are many other more pressing things on our plates. What we have to deal with now is the situation. And a situation which, by the way, is of express danger to national security." Richards cleared his throat. "There is a possibility that Scarborough is in league with KGB subterfuge."

Myers's eyebrows rose. He stared at Richards for a moment and then chuckled softly. "Whew. And you're expecting me to buy all this?"

Richards frowned. "These are the *facts*, Myers."

"Okay, so maybe I'm too exhausted to play E. Howard Hunt to your G. Gordon Liddy, *sir*. I apologize. But Scarborough couldn't do this kind of stuff. I know him well enough for that . . . And what kind of national security are you talking about? He's a UFO skeptic. That's a cold topic." Myers heaved a sigh, and rubbed his eyes. "Besides, we haven't gotten to the nub of the matter yet, have we? Why are you telling me this? Because I know the man? You want my opinion? Well, my opinion is that Dr. Everett Scarborough is a great guy and all this must be some kind of dreadful misunderstanding."

Richards pushed the tape deck forward. "This is why I called you in, Ed." He hit the "Play" button.

Everett Scarborough's voice, sounding uncharacteristically tense and uncertain, came through the speaker, addressing Ed Myers. He listened to the tape. There were three messages, the last two just short inquiries, both promising to call back.

Richards turned the tape off.

Myers sat for a moment, absorbing this. Then he turned and looked at Richards. "You bastards," he whispered.

"Name of the game, Ed. You're one of us. You know that these measures are sometimes necessary."

"That's *my* private line! I'm high enough for the privilege of *not* being bugged."

"No bug. We just checked your tape. We thought that Scarborough might try to get to you for help. We turned out to be correct, Ed. And so, *that* is why you're here. Tell me, would you have aided a wanted fugitive?"

"It might be nice to *talk* to him. To see *his* side of the story. There's no law against that. That's certainly far short of treason."

"Certainly. In fact, that's just what we want you to do, Ed. We want you to talk to Everett Scarborough. We want you to persuade him to turn himself in, so that this whole sorry business can be resolved."

"And if he won't turn himself in?"

"We have a contingency plan for that."

"You mean that you want me to betray a *friend*!"

"When we take our oaths to this agency, and to our country, Ed, we take on a difficult responsibility that some times necessitates actions we loathe. Does the trashman relish rotting garbage? I think not—but it's his job. This is your job, Ed, but more than that, it's your duty." Richards's voice and posture assumed a standard officiousness, edged with a steely, cutting nasality peculiar to him in times when he exercised his authority, his command. "Need I remind you of those oaths you have taken, to say nothing of your responsibilities as a citizen of the United States of America? Need I remind you of the ample financial rewards you have received in line of duty." Richards sighed deeply, looked away, then idly thumbed a button. The drapes concealing the plate-glass window softly shushed apart, like the dresses of a woman, revealing the soft spring hues like watercolors on the palette of Virginia forest. "Nice view, huh?"

"Lovely." Monotone.

"You think I didn't have to stab some people in the back to get it?"

"God alone knows what you've done, Richards."

"What's that supposed to mean?"

"Nothing."

"You've got, what, two, maybe three years before you're pegged for a nice desk job. No more gallivanting around the world, eh? Still, I suppose you're not exactly relishing the idea. Desk jobs can be dead ends, can't they, Ed? You know that; I know that . . . I could help you get a view like that." He gestured out the window. "And the power that view brings along."

"Jesus Christ, Richards. Jesus fucking Christ, and I'm wasting what's left of my soul, taking the Lord's name in vain on account of the like of scum like you."

"I realize that I've put you under some strain, Ed. I'll take that into account. I'll ignore the name calling."

Myers covered his face with his hands, breathed, took in the carbon dioxide. "Nope. Fry my ass. Give me a nice lowly desk job. I don't care, Richards. I know other super-grades. You're not the only bigwig around here. Something will look mighty fishy if you start dumping on *me*. To paraphrase a recent esteemed member of our damned corps, *Read my lips*. I will not do anything to betray my friend Everett Scarborough."

"Oh, but you must, Ed. You see, you're the only one within our number who *can*." Richards tented his fingers a moment, thinking, then opened a drawer. He pulled three files from his desk. One was blue, one was white, the last black. "Edward Myers, you have been privileged before to see classified materials. That is part of your job. However, as a lower grade, you of course are not entitled to see certain reports; you are not privileged to know certain truths. I think it's about time, considering your extreme importance to us, to let you know a few facts. In these folders, more or less, are the reasons we need to contain the threat known as Everett Scarborough."

Enlivened a bit from his exhaustion, clearly intrigued but loathe to show it, Myers looked down at the folders.

"Fancy," he said.

"Project Blue Book. Project White Book." Richards reached out, took the corner of the black folder and pulled it away. "Project Black Book."

"I should have guessed. UFOs. So what. Everybody knows that the government still secretly monitors reports. Why all the show business, Richards?"

"Just have a look. I've prepared them especially for you."

Myers picked up the folders. He paged through material of the blue, then hurried onto the white. He read for several silent minutes, then looked up at Brian Richards as though he were Satan himself.

"Hey—wait a second. Christ, Richards . . . This . . . This . . ." He couldn't get out the words.

"You've seen it now. You didn't have to look, my friend. If you're with us, you'll not only have all the things I promised, but also a hefty bonus. A hundred thousand. No, let say two hundred thousand dollars."

"This *can't* be authorized, Richards! *This* is treason! Treason to the Bill of Rights, to everything that is holy, to human *decency*, for God's sake."

"It's all because of this," Richards waved the black folder tantalizingly. "If not for Black Book, the other two would not exist. Here, you might as well read it."

"Why?"

Richards tossed the folder on top of the others. "Ed, we're down to the bottom line here. It looks as though we're in a bit of a bind, here. I have no choice."

"No choice. What do you mean?"

"What time is it, Ed?"

Myers checked his watch. "I've got 2:32 P.M. Eastern Daylight Savings Time."

"How precise. Let me make this simple. If you do not now agree to help us to capture Everett Scarborough, by 2:32 P.M. tomorrow, your family will have experienced unfortunate and most tragic accidents. *Fatal* accidents. Now. What would you prefer, Ed. A dead family and a loyalty preserved, or a little help bringing in a criminal?"

Ed Myers went white. "This isn't just Company talk."

"No."

"I've heard whispers."

"Probably true, Ed."

"You're one—"

"That's right."

"Oh, God help me."

"I doubt it, Ed." Richards went to the private bar by the sofa and fixed Ed Myers a stiff drink. He knew that Myers liked dark rum and Cokes, and it took just a moment to fix one. He went to the ashen man and placed the drink in his hand. Myers gulped half of it down, his knuckles turning white on the arm of the chair.

This was quite typical.

Richards had experienced three, maybe four situations, quite similar to this. Sometimes, the Publishers had to recruit on sudden notice, and as Editor-in-Chief now, Brian Richards had done a few himself, mostly fellow CIA operatives.

They'd all reacted much the same as Ed Myers was reacting.

"Maybe you'd better read the Black, Ed. You'll understand more then."

Ed looked at the Black folder. His hand shaking slightly, he opened it and started to look at the papers it contained.

Casually, Brian Richards went back to the bar, grabbed the bottle of Jamaican dark rum and liberally freshened his new recruit's drink.

CHAPTER FIFTEEN

The standard route from New York City to Baltimore is simple. You just take Route 95 south. This involves traveling much of the trip on the dreaded and smelly New Jersey Turnpike, crossing the majestic Delaware Memorial Bridge, and then paying another two dollars' worth of tolls to finish up with the JFK Expressway. The total trip takes between three and four hours, depending upon which side of the speed limit you care to drive.

Everett Scarborough, however, elected to take the long route. He slept in a Motor Court Lodge in Elizabeth, New Jersey, catching up on his rest. Then rising late in the morning, he took side roads down the rest of way, stopping in Wilmington, Delaware, to call directory assistance, and then Mrs. Elaine Strazinski.

Mrs. Strazinski took some convincing, but he finally convinced her he was a legitimate UFO researcher, driving through Baltimore. Could he interview her this evening?

Yes, she'd said. That would be fine.

Scarborough arrived in Baltimore at seven o'clock. His appointment was at eight, so he had time to find a hotel for the night and pick up Mrs. Strazinski's order. He found just the sort of motel he was looking for in Dundalk, near the Maryland Port Administration. Shabby. Cash only. Quiet and indifferent. Then he found a liquor store, purchased a case of National Bohemian beer, and drove the remaining miles to the portion of Baltimore called Highlandtown.

Baltimore is, generally speaking, a working-class town, with many working-class neighborhoods. When Mayor Schaefer dredged the harbor and created touristy Harbor Place, the "Renaissance" he erected, although impressive, did little to change the working-class sections like Highlandtown. The area was famous for its Polish population and infamous for the sexual profligacy of its young unmarried women.

The neighborhood itself was a series of modest, tacky-looking rowhouses. One of the things that always impressed Scarborough about Baltimore was the lack of trees, along with the general, depressing external feel of neighborhoods like Highlandtown. They had a dull and shrunken look, few trees, and the general air if not of poverty, then certainly nothing near affluence, either.

Scarborough found a parking spot for his car easily enough on Mrs. Strazinski's street; 9427 Lombard was the address. The smell of tangy tomato sauce, cheese, and spices drifted from a pizza shop on the corner, making Scarborough realize that he was hungry. He would have gone in for a slice, but he remembered that Mrs. Strazinski had offered him a sandwich. It was best, he found, on such interviews to accept food. Maybe it made the interviewer feel more like family, Scarborough wasn't sure—but it certainly seemed to make the interviewees more comfortable. Scarborough found 9427 easily enough, and walked up creaky wooden steps. The button he pushed by the door failed to ring a buzzer or a bell—so he knocked. He held the case of cold bottles by his side.

A young boy answered.

"Hullo?"

"Hi there. I'm here to see Mrs. Elaine Strazinski? My name is Ted Anderson."

"I'm Billy."

"Hi, Billy."

"You bring anything for *me*?" said the boy, eyeing the case of beer. Scarborough figured he was eight or nine years old.

"No—wait a minute. How about a quarter!"

"Geez! You can't buy shit for a quarter these days."

Scarborough smiled. "No. Inflation and all that. How about a dollar bill."

The boy grinned. "Sure." He turned and hollered back into the house. "Mommmmmmmmmmmm! There's a guy here to see you."

"Just a minute . . ." called a voice. "Tell him to come in!"

"Come on in, mister." The boy held the screen door back for Scarborough, allowing him to step in.

Inside, the house smelled of cabbage, cats, dirty laundry, and Airwick. Billy accepted the dollar bill that Scarborough dredged out of his pocket, then led him into the living room. An overweight man sat in a tattered Lay-Z-Boy recliner, a twenty-five-inch RCA television flickering and chattering a baseball game at him. The man seemed semicomatose, holding a National Bohemian can in a Styrofoam cooler marked Ocean City, a battered Lay's potato chip bag in his lap. The man wore a sleeveless T-shirt and no shoes.

"Hello," said Scarborough. "You must be Mr. Strazinski."

The man grunted and glanced at the TV screen. "O's," he said. He belched, and brought the can of beer to his lips. Sipped.

"Yes, of course. The Orioles. Hope they're doing better this year than last."

The burly man glared at Scarborough a moment, then turned back to the TV set. He reached down and scratched his crotch, then returned his attention to the game.

"Would you like a beer?" asked Scarborough. "Your wife said to bring along a case. I'm Ted Anderson, by the way. I'm here to talk to her."

From upstairs, there came the screech of a young child, followed by motherly cursing. Mr. Strazinski did not seem to notice. He sloshed his beer for a moment, found it wanting, put it down, and accepted a bottle from Scarborough.

"Thanks." He nodded toward a seat, and Scarborough sat down, propping the case of beer on his knees.

He opened a bottle of beer for himself and watched the game. The beer, though cold, was thin and bland, with a nasty aftertaste. But he was thirsty, so he drank it.

The place looked as though the Strazinskis attempted to keep it neat, but were failing, due to the rapid entropic effects of too little money and too many children. Above the sofa, displayed proudly, were two large portraits of large-eyed, sad children, each holding out flowers. On a shelf by a record player was a small statuette of Elvis Presley in his Vegas years, marked The King Lives! Scarborough noticed that the tatty chair he sat in was covered with white cat-hairs. He took a few more gulps of beer, praying that his allergies would not act up too badly.

He had almost finished the beer, when a woman walked into the living room. "Mr. Anderson! Well, hell-o! I'm Elaine Strazinski! Pleased to make your acquaintance." She held out a hand for him to shake and Scarborough shook it.

Elaine Strazinski wore a bright yellow and red dress with flounces on the sleeve. Her hair looked newly permed, and she had on an abundance of makeup. Neither the dress nor the makeup hid the painful fact that she was a fat woman in her midthirties who had perhaps been pretty once, but now looked mostly used up in a way that lower-middle-class women with three or more children tended toward.

"Thank you for letting me come over on such short notice, Mrs. Strazinski. I hope I haven't inconvenienced you."

"No, hon, of *course* not. I was wondering when one of you

investigators would get a sniff at that story and hightail it on down here to Baltimore!'' (She pronounced it ''Bulmer,'' in the classic local fashion.) ''Say, I see you got that case of Boh, and you've already helped yourself to one. Well, good. And bottles. You see, Earl, this is a man who knows his beer. Bottles always taste better than cans!''

Earl grunted.

''Don't mind Earl. He's just put in a long day at the muffler shop. Hell, if I had to get up under cars all day I guess I wouldn't be much good at night, either. Why don't you just carry it on back here to the kitchen and put it in the fridge and we'll just sit down awhile and talk.''

Scarborough picked up the case of beer and carried it back through the long, narrow railroad-car house. He had to avoid a skateboard and a batch of marbles, and a cat rubbing up against his legs as he walked through the dining room, but finally he made it safely back to the kitchen.

''Right in here, Mr. Anderson,'' said Elaine Strazinski, holding open the door of a late-model Frigidaire. ''Bottom shelf. Pick out one for me, and get yourself another. I got a plate of crabcake sandwiches fixed up for you. Figured you come to Bulmer, you might as well eat a local delicacy.''

''Thanks very much, Mrs. Strazinski.''

''Elaine. Just call me Elaine, hon. And I'll call you Ted.''

Scarborough took another beer and sat down on a chair by a sparkling Formica table. He could see why Elaine Strazinski had brought him back here. It was the nicest room in the house, equipped with new cabinets, bright linoleum floors kept spotlessly clean. Large, it had all the trimmings of a modern House Beautiful, from dishwasher to microwave oven. Its window looked out onto a narrow but long backyard filled with kids' toys and bikes, and a leaning barbecue cooker. The backyard was shadowed with dusk.

''I figure this is the place I spend most of my time vertical in this house. Might as well be the place I put that money I got from that book that mentioned me,'' Mrs. Strazinski said.

''Ah. That would be *Hijacked by the Stars*, wouldn't it?''

''You bet.'' Elaine Strazinski pulled out a plate of cold crabcakes, a jar of Gulden's mustard, and a loaf of Wonder Bread. ''That Mr. Abe Dickens, he gave me a little cut of the royalties for talking to him so long. My story took up half the book!''

Abraham Dickens was the chief competition to Budd Hopkins for third-person UFO-abduction books. The only thing that held the books together was Dickens's extraordinary gift for sincerity even as he feverishly described the most frightening and unlikely situation, creating a far more spiritual but just as scary overview of the phenomenon. In one of his own books, Scarborough had described him as ''the George Adamski of the paranoid eighties and nineties.'' It was Whitley Strieber who had suggested that if the alien-abduction archetype was not ''real'' in terms of everyday reality, then it could be the beginning of a new religion. If this was so, then Abraham Dickens was the John the Baptist of that religion.

''I can't help but be curious, Elaine—why isn't Mr. Dickens taking this story?''

''Oh, he'll get it all right. You think I'm stupid? Shoot, who knows, maybe I'll get in another book. No, Abe's off in Australia, doing a book called *Dreamtime Abductions*. Real metter—metter . . .''

''Metaphysical.''

''Yeah. That's the word. Anyway, he's a nice, sincere guy, and we don't have no exclusive contract or anything.''

''You know, I can only pay maybe a couple hundred dollars . . .''

''Don't worry, mister. I figure if I charge, people'll think I'm making this up. And I swear to God, I ain't! Abe didn't give me nothing for my story at first. No, mister . . . I just gotta tell my story. I consider it a public service! You know, I been on Oprah and Donahue and Geraldo and even Sally's little show. . . and I'll do more if I can. People gotta be warned about what's goin' on.''

''Yes, of course.'' Scarborough found himself biting his tongue to prevent the usual phrases of disbelief, sarcasm, and witty derision from escaping reflexively. Actually, he had an entirely different opinion now about the alien-abduction phenomenon. Now, he was at least open to *listening*.

''So you're the first one who wanted an in-depth interview on the story I sent in.''

Elaine Strazinski made him a crabcake sandwich with lettuce, tomato, and mustard. She cut it in half, put it on her best china, and placed it in front of him like an offering it pleased her to make. In fact, it was delicious, and Scarborough said so.

She beamed. "Thank you. Say, you know, I was just thinking . . . you look kinda familiar. Ain't I seen you somewhere before?"

Scarborough cringed. Oh God, he thought. He hoped she didn't have any of his books. Especially the hardbacks, the ones with his picture on it.

"That fella that was in those shark movies . . . Yeah, *Jaws*!"

"You mean, Roy Scheider?"

"Yeah. You look a little bit like him, only you're scruffier."

Scarborough shrugged, relieved. "Some people say that, yes. No, I guess you'd called me a freelancer. I do pieces for lots of magazines and newspapers. I like to travel you see—it's a very free kind of life."

"Well, good for you. I always wanted to travel when I was a girl. See the world." The woman sighed, looking a little wistful. "Course you can't do that when you've got a husband and four kids. Little darlins are tucked away in their beds now, all 'cept Billy. You met him. Took hell and high water to get 'em in there this early, but I did."

Scarborough took the clipping from his pocket and smoothed it out on the table. He took out a pair of glasses, put them on—and then opened a notebook to a blank page, adding an extra weight of professionalism to every movement.

"Now then . . . Elaine. You of course have told your story in the Dickens book—and on TV. Let me see if I've got the basic facts correct. Camping trips to Pennsylvania. Abductions from your tent. Confrontation with aliens. Examination. Experimentation. And later, they actually came here, to your house. At that time, several years after the first abduction, they displayed a hybrid child they said was yours. Now, this is similar to other women's experiences, especially to that described in *Intruders*, by Budd Hopkins. The difference being, as I understand, that on the flying saucer you met Jesus Christ, the Buddha, and Mohammed, all of whom reassured you, promising that your lost child would be a member of some new star race." Scarborough spoke in a monotone, without a trace of judgment in his tone.

Elaine Strazinski grabbed a handful of tissues from a daisy-covered Kleenex box and dabbed at her eyes melodramatically. "Yes, yes. God, that still hurts . . . losing my baby. I've got four here and one in the stars, Ted. I love 'em all. Sometimes I wake up in the middle of the night, yearnin' for Zeta . . . That's his name. Zeta. Such a golden little boy. He shone, Ted, like there was this halo, all around his body. What they do to you, Ted . . . it ain't right. Only way I figure, they just don't understand, those aliens."

"No. It's not right."

"Still and all, my Lord Jesus told me it was okay. I don't know what those other heathen were doin' there on that saucer, but maybe that's just God's way of saying to me, Elaine . . . you oughta be more economical!"

"Uhmm—you mean, ecumenical, don't you?"

"Yeah, like that . . ."

"Have your husband and your kids seen these aliens?"

"Billy has. But the others . . . they know something funny's goin' on. Earl, he tried to just ignore it. It's too much for him. He calls it my little hobby, and he doesn't like to talk about it. Maybe that's why I like to talk about it to other people. Get it off my chest, you know."

"Yes, of course. But now, this very short piece appeared a few days ago. And all it says is that you were taken again. Only this time, you saw a human being on board the saucer. A very beautiful woman, you say. Quite mortal. Definitely not the Virgin Mary." It took every fiber of concentration in Scarborough not to sound ironic. "This, as far as I can tell, is the first case of an individual seeing another human aboard one of these craft. The usual situation involves a single human and a number of aliens. As I recall, your reports of the physical descriptions of these aliens are quite similar to other reports—short creatures, triangular faces, almond-shaped eyes almost glittering . . . But this episode is quite unusual. Could you tell me about it in more detail? I think that a more extensive article might be called for."

Clearly he'd said the right thing. Plainly, this kind of attention was exactly what Elaine Strazinski was looking for. Her whole aspect changed. Suddenly she was "on"—as though she were speaking in front of an audience or on the Oprah Winfrey show again. Her gestures grew broader, her eyes became wider—her breathing quickened.

Scarborough could not help but feel a sudden despondency fall on him. Everything so far had about Elaine Strazinski was *exactly* conforming to the outline of the "Typical Saucer Slut" in *Above Us Only Sky*, a particularly controversial chapter that he'd regretted slightly, and was thinking of taking out of the book ("Or at least tone down the sexism some, huh, Dad?" Diane had suggested.) for the paperback edition. Neglected middle-class housewives looking for sexual, spiritual, and certainly psychological relief, brewing up hypnotic-induced delusions and fantasies based on suggested dreams. Attention was what they most lusted for . . . attention, and excitement that, before their experience, their dreary and troubled lives so lacked. You only had to take a ten-minute listen to the chubby, distressed women that Budd Hopkins trotted around on the talk shows to publicize *Intruders* to see that they weren't quite right in the head, and chances were their psychological difficulties dated back way before any run-ins with the saucer-folks.

Nonetheless, Scarborough listened, trying to rid his mind of prejudice as best he could.

"You know, since the book came out, they simply stopped coming to get me. I mean, we even went camping once 'cos I wanted to see if they'd pick me up. And the bastards didn't! I thought maybe they were peeved that I'd told the story and lots of people knew it, even though they told me that it was okay to do that; and the time I saw Jesus, he said to spread the Gospel. Gospel means 'good news,' so I guess He wanted me to tell about the saucer people, so I did. So this is why all this comes out of the blue."

"Were you camping when *this* incident happened?"

"Nope. Little too cold at nights in the Poconos for that. No, I'd just fallen asleep on the couch down here. Next thing I know, there they are. The saucer people. They start to cart me away, and then I went out like a light. I wake up again—and I'm in the saucer. Well, I say it's a saucer, 'cos of what I seen in the Poconos, that sighting I made before all this happened."

"Yes. I remember." Not the truth, but Scarborough could pretty much guess what it was supposed to be. UFOs and saucers came in all shapes and sizes, but they basically either floated or zipped about above the ground, glowing with different lights.

"Okay. So I think—hey! Homecoming! Maybe I'll see my son again. That's not so bad. Like, I ain't saying that getting carted off by these bozos is a load of laughs. It's pretty damned scary. But I was almost starting to miss the change of pace. The 'Lay-Z-Boy' that belches out there isn't exactly Mr. Excitement. This was kinda different though . . . I could tell almost from the very start."

"Why was it different, Elaine?"

"The inside of the saucer. For one thing, it looked a little different. Had a different feel to it. Hell, *I* felt different. Like, I was on a different prescription or something."

"Drugged. You think you were drugged all these times."

"Well, of course. Or, like Abe says, the equivalent. The mess with the neero-chemicals, that's what Abe says. Could be like a electro-magneto spectrum gun or something . . . Anyway, it's alien, and it puts you out. How come you think Earl and the kids stay out while little creatures float me out of the tent—or out of my own house, for God's sake."

"Yes, of course. I'm just trying to get this in some kind of specific detail."

"Okay. So, they do the usual checkup. That's never fun, and I complain, but what are you gonna do?"

"Checkup. Could you be more specific?"

"It's a little embarrassing. Let me give you a paperback of Abe's book."

"Oh. You mean, like amniocentesis. Blood sample. Some gynecological stuff."

"All those words, yeah, but it just seemed like a formality. So I ask them, are you going to make me pregnant again? And they say, no. And I say, good. What the hell do you want with me now? Haven't you bothered me enough. I figure, I don't raise a fuss, they'll think I'm dead or something. And that's when it happens."

"The human woman came in?"

"No. There I am in the middle of that alien examination room. You know, lights and doodads all over. Real strange stuff, like that movie *Alien*."

"The artist Giger . . ."

"Whatever. So, suddenly everything gets all rushy and black, and its like I had a couple shots too many down the block at Hardigan's Bar. I go down, and things, whew, they start spinning like crazy. And the aliens . . . they just freaked. And then things started to go off and on—"

"Strobing."

"Yeah, like special effects for a rock group I seen once down at Hammerjack's. And the aliens, they leave and I'm down on the floor. And the lights come back on and I'm looking up at a woman."

"A human woman."

"Yes."

"Elaine. Now, this is very important. Could you describe the woman to me? That part wasn't in the article."

"Sure. Only the memory's kind of patchy. She was really beautiful, but she wasn't wearing any makeup."

"Was she young?"

"No, but she wasn't old either. I'd say maybe her thirties."

Scarborough's heart sank. No, it wasn't Diane. However, he knew that chances were it wouldn't be anyway—this was a long shot. Still, he wanted to hear the woman out. This could be significant.

"I see. Try hard to remember what she looked like . . . and what she was doing?"

"Doing? Well, she looked *real* upset. She was shouting stuff. And I remember stainless steel-like instruments coming down. I felt a sting on my arm on a different place than usual. Here, let me show you."

She rolled up her sleeve and showed Scarborough a triangular scar on her arm, much like the stigma described by other victims of alien abduction. It looked like a hickey from a cookie-cutter machine, thought Scarborough.

"Ah yes, of course. But the woman. What did she look like?"

"You know that show 'Cheers.' The one in the bar."

"Yes, of course."

"That woman who married the shrink. The one who acts like she's got a broomstick stuck up her ass."

"Lilith. Lilith Sternan. She married Frazier Crane."

"Yeah, so you're a 'Cheers' fan, huh? Well, this woman looked like her, only she had long blonde hair tied up behind her head. Made her look real severe. And glasses. She wore glasses too. She was saying something about 'the mix.' Something was wrong with the mix and if they didn't get some kind of chemical in fast, they'd have a corpse on their hands."

"And you remember this—"

"Sorta. I went under for a little bit, and then the aliens came back. They told me that the woman I'd seen was one of the other earth women they were experimenting with. Besides, I wouldn't remember this anyway. What I was supposed to remember was that this wasn't a fantasy, it was really happening—and you know, spread the gospel. But the Lord Jesus wasn't there. Nope, and no sign of those heathen messiahs either."

"And when you woke up—"

"I was back on the couch, with this goddamned hole in my arm! Shaking and crying . . . And you know, after I settled down, I went up to Earl, and he was sleeping like the dead. Same with the kids. I tell you, they do *somethin'* weird to you, all right." She frowned and shook her head sadly.

"I see. You know, this could be very significant."

"Could it?" She brightened.

"Yes. I definitely think so, Elaine. Did you have any kind of feeling that this woman was in *charge* of things?"

"Yeah, you bet! That's what I told the newspapers I called. I said, you know, I think that there are *humans* involved with this stuff, somehow. I couldn't figure out . . . Maybe the human race has turncoats. That's what Abe thinks, anyway. You know, that there might be people the aliens use to help kidnap people. On the other hand, there's also another possibility."

"What's that, Elaine?"

Elaine Strazinski's eyes got wide and wild. "Maybe that's one of the things they're doin'. Maybe they're making *artificial* human beings. Things that look just like you and me and walk among us, only they *ain't* human. They're one of *them*!"

"I see. Well, I'll be certain to include this in the article. Thanks so much for your time, Elaine. I won't keep you any longer."

Elaine blinked. "Say! You're not going yet. You haven't finished your sandwich. And why don't you sit and have another beer? You know, Ted, you don't look so good. I think you need some food and another brew." Suddenly, she was up again, flouncing about, turning on her femininity and fussing over him, going to the Frigidaire and pulling out a beer, unscrewing and tendering it toward him as though it were some sort of offering.

Scarborough thought about this a moment.

Although he hadn't gotten what he'd hoped for, he'd picked up intriguing evidence. Assuming of course that all of this wasn't just a part of some psycho-fantasy, part of the overall delusional network, as he had dubbed it in his books. Assuming that Elaine Strazinski *had* been actually kidnapped . . .

Then, chances were, the abductors hadn't been aliens.

That was just the thing about the alien abductions, and Scarborough had fallen into this trap as well. The key word that attracted attention was *alien*. Abductions happened everyday, the FBI had a whole department devoted to kidnapping and such. But the concept of "alien" polarized listeners to "yea" and "nay" camps. The believers bought the story, mostly because they were open to the idea of ETs anyway. The skeptics, like Scarborough, had discounted the stories entirely because they did not see any logical thread, nor did they believe that extraterrestrial intelligence was visiting earth.

But what if these abductions—at least some of them—were real, but not necessarily alien. What if either they were interpreted as being alien . . . *Or they were made to look alien*!

A month ago, Everett Scarborough would have discounted all that as poppycock.

Now he was not so sure.

He hadn't gone through the hell of the past two weeks back then. He hadn't realized that pure paranoia *could* be reality.

He'd interviewed abductees before, many of them quite like Elaine Strazinski in their convictions, many of them just as vocal—people who passed on the word to others. Some abductees were clearly the wackos and maladjusted that Scarborough had interpreted Elaine as being at first, profoundly influenced by the "Typhoid Mary" types like Whitley Strieber, Maximillian Schroeder and Elaine Strazinski.

Only *they* hadn't described to Scarborough an encounter with a human female—the description of which sounded like someone that Everett Scarborough *knew*. Scarborough accepted the beer and the offer of another sandwich.

Maybe there were other things that would come to this woman in another half hour of talking.

So, for the first time, as he ate his crabcake sandwich and drank his National Bohemian beer, Everett Scarborough just sat with a victim of alien abduction and *listened*.

CHAPTER SIXTEEN

"Hello."

"Ed? Ed Myers?"

"Yes, that's right. Who's this?"

"Ed, it's Everett. Everett Scarborough. You're back! I've been trying to reach you."

"Jesus. Ev! Are you okay, man? I just got back from a trip today, and my wife told me . . ."

"Yeah, Ed. I'm in deep trouble."

"Where are you?"

"I can't say."

"No. No, of course not. I'm just glad to be here for your call. You know, I haven't even had time to play back my phone machine. You must have been calling me a lot."

"Yeah, Ed. I have to call in a favor."

"What can I do to help, Ev?"

"First, do you believe that I'm innocent? That I didn't kill Mac or that CIA agent?"

"Ev, I *know* you. Why would you do a thing like that. And I know my Company. They'd set you up in a snap. It's a frame, Ev. It's gotta be."

"Oh yes. A frame. But more, it's more than that. Some outlaw group of your people have Diane. I don't know why they haven't used her against me, threatening to kill her unless I turn myself in . . . But maybe they're being cagey. I don't know. Ed, there's something incredible going on. Some kind of clandestine conspiracy, the likes of which I'd never *imagined* before. I can't even begin to guess why it exists, but it involves the whole UFO business, all the way back to Project Sign in the forties."

"I see."

"Look, I can't talk long. I know you've never been real thrilled with your employers, but I'm not going to ask you to betray them. I just need some information. Information that I think you might be able to get access to. And I'll take it from there."

"Gee, Ev. I don't know... There are all kinds of levels of clearances, and I'm far from the top."

"Yes, but you've got your methods."

"True. Very true. But Ev, maybe we better get together and talk."

"No. The phone will do for now, which is why I can't talk long. Maybe later. Right now, I want you to take down these names. Colonel Walter Dolan. Brian Richards. An organization called the 'Editors and Publishers.' And lastly, see what you can get on a woman named Doctor Julia Cunningham."

"Right. Got them. I'll see what I can do."

"Thanks, Ed. God, it's good to hear your voice. You don't know what I've been through."

"Look, has it occurred to you, Ev, that you're not thinking straight. Maybe we should get together. I can fly anywhere you want—I've got lots of Frequent Flyer Miles, and some personal leave to burn up..."

"No. We'll talk about this later. I can't talk any longer, Ed. I'll call this time tomorrow night, if you can be by your personal phone line then."

"You bet, pal. And I'll see what I can come with."

"Thanks, Ed. You're a lifesaver. I wish I could just sit down and have a beer with you. I'm just glad that there's someone in the governmental loop I know I can trust."

"Talk to you tomorrow night."

Scarborough hung up the public phone outside the 7-Eleven store by Fayette Street. He went back to his car in the lot, and drove the rest of the way down South Broadway, to an area of Baltimore he and Mac MacKenzie used to go to sometimes, called Fells Point. He felt like having a good import draft beer or two to get the taste of the National Boh out of his mouth, and Highlandtown had been pretty close, so he figured that since all he had was a tatty, depressing room waiting for him back in Dundalk, he might as well take an hour or so and visit Bertha's, or maybe the Wharf Rat. If he was hungry later, he could get some of that good Mayan Mex food at Mike's Tavern.

Scarborough was feeling buoyed by the phone call. He'd made it through to Ed Myers. The conversation had been short and cooperative. Ed hadn't tried to keep him on the line, which reassured him immensely. They'd talked too briefly for anyone to get a tracer on the origin of his call; the fact that Ed had let him go so quickly meant that Ed was being truthful in his willingness to help out. This relieved Scarborough's paranoia greatly.

He'd tried to call Marsha earlier, but had gotten nothing.

He'd try later—or maybe tomorrow night.

What next? Maybe a couple of beers would help him decide.

Fells Point was a historic port area jutting into Baltimore Harbor like a stubby finger. The place once had a reputation for wild bars, derelicts, and sailors aprowl on leave. It still had a high density of bars, but in the last decade or so, the "Renaissance" of Baltimore had led to extraordinary gentrification that upped property values and made Fells Point an aspiree for the title of the Georgetown of Baltimore. Fortunately, despite the coming of the upper-scale shops and restaurants, the area still maintained its cluttered bar flavor, along with its reputation as a good place to go see folk or blues or jazz on weekends.

On Fridays and Saturdays, a jazz group called the Alan Houser Quartet played at the bar of Bertha's, a restaurant famous for its steamed mussels. Mac had liked the band a lot, because they mostly played bebop jazz. Scarborough preferred cool jazz and swing himself, but he liked bebop as well, and whenever Mac came in from Iowa for a visit, they tried to make the trip. This wasn't Friday or Saturday and Mac Mackenzie was dead, but it was still Fells Point, and Scarborough needed a pint of Newcastle Brown Ale.

He found a parking place by the marketplace situated on the wide island that divides the entirety of the two lanes of South

Broadway, and walked past Max's and then across a parking lot toward the colorful signpost displaying the name Bertha's. To his left a half-moon rode over the Harbor like a memory of earlier, sailors' days in Fells Point, when merchant ships and naval vessels would dock all the time, their seamen spilling into the seedy bars and whorehouses that lined the streets. This, after all, was the area where Edgar Allan Poe was supposed to have had his last drink.

As he headed toward the bar, the hackles on Scarborough's neck rose. A wave of gooseflesh passed over his body. He stopped in his tracks and swivelled around.

Just the usual evening bar-hoppers, restaurant patrons, and theatergoers. No one seemed to be paying the slightest amount of attention to a scruffily bearded guy in flannel shirt, windbreaker, and jeans. Still, Scarborough had learned to pay attention to these premonitions. He even considered cancelling the trip to the bar, and heading back to Dundalk. Just then, though, the old red door of the Bertha's opened, letting go a breath of Stan Getz doing "Old Copenhagen," and Scarborough lost his hesitancy.

Bertha's bar was a long and narrow affair, dark, with smudged windows. It smelled of spilled beer and cider and the delicious fish chowder and scotch eggs it served as bar food. All manner of peculiar signs were displayed on the walls, and from the rafters hung old musical instruments like somnolent bats. It was quiet that night, with only about ten patrons, four of whom were sitting at the two booths. An attractive brunette with a sweet Irish smile was tending bar. She brought him a delightfully dark English mug of ale for only two-and-a-half dollars, and Scarborough sipped at it gratefully. Too cold. British drafts should be served at room temperature. But delicious, nonetheless. Anyway, Newcastle was famous for its potency, so Scarborough didn't intend to drink more than one, and he was going to let this one linger—so by the time he was half-through, it would be about right.

He quietly sipped his drink, listening. Getz melded into some more upbeat Dave Brubeck and Paul Desmond, and then settled down into some classic Zoot Sims and Al Cohn. Ah, sing, sweet saxophones! Already, the National Boh taste was gone and his nerves had at least stopped their dissonant screechings.

He closed his eyes and listened to the pleasant unobtrusive jazz. Goodness, you could *feel* the ambience of this place even with your eyes closed. Scarborough was glad he had come. It was probably close to the meditation Diane was always after him to do. He could almost hear her plaintive voice in his ear now: "But Dad, you've just *got* to find a way to wind down without drinking alcohol and listening to this silly, dead music!"

Scarborough sighed.

He'd give up drinking. He'd give up jazz. He'd give up just about anything, if he could only get his daughter back, away from those heartless monsters in the grey suits . . .

By the time Scarborough finished his ale, he'd learned the bartender's name was Ivy and that she was a sculptress and had a show coming up at a local gallery. He gave her a nice tip and exited regretfully. Time to go. He'd about had his limit, and he didn't want to get pulled over by a cop for a little Breathalyzer test.

As he stepped out onto the sidewalk, he caught something in his peripheral vision.

A man disappeared behind the doorway of the market.

Or it looked like a man, anyway. Had it been his imagination?

Was someone following him?

One of the two men in the subway station?

Scarborough had to find out. He knew that if he started walking toward the dark doorway, the man could scrabble away easily. So, the thing to do was something unpredictable, to draw the man after him.

He turned down Shakespeare Street, and stalked quickly away.

Shakespeare Street was narrow and paved with cobblestones. The fixtures in the lamppost were new and bright, with an artificial Victorian feel to them—all a part of the new upscale look to the street. Scarborough walked away, then turned right on Dallas, making it look as though he were headed for Lancaster or Aliceanna on his way to Little Italy.

Instead, however, he stopped and retreated into a garage doorway. The turn onto Dallas had taken him out of New Yuppieville into Old Industrial Baltimore, a bleak cityscape of warehouses and old tired-looking townhouses, without a tree or patch of grass to be seen anywhere. The area was deserted, except for the occasional car.

Scarborough waited, scarcely breathing.

Nothing.

Had he been wrong? Was his paranoia acting up again?

He was about ready to give up, take a deep breath, and head back for his car, when he heard the footsteps.

Click, shuffle . . . click, shuffle . . . Wary footsteps, but footsteps that were in a hurry as well.

Scarborough flung himself back into the doorway and got as quiet as he could, fast.

The footsteps came around the corner and turned into a man. The man wore a leather jacket, jeans, and tennis shoes. He could have been any other Baltimorean walking the streets away from Fells Point. Only, from the way his head turned, from the way his eyes moved, it was clear that this man was *looking* for something or somebody.

And that *feeling* was starting to percolate in Scarborough's gut, that terrible something he'd tried to ignore.

He waited for the man to pass, and then jumped him.

Scarborough was the bigger of the two, and his weight drove the man down. The man's head hit a trash can. It banged, rolled away, felled another trash can. The man struggled with surprising strength and agility, but Scarborough was able to roll him over and get a look at the face.

It was one of the men from Tower Records in New York.

Not the older man.

The younger man.

Scarborough hit him in the abdomen, hard, and air whooshed out the man's mouth. Scarborough grabbed his neck and started squeezing. He had the strength of a madman. All his frustration, fear, and worry were like coiled springs in his body, releasing.

"Who are you?"

The man just gurgled and choked, strong hands clawing at Scarborough's windbreaker.

"Who are you? Why are you following me? Where's your friend?"

The man just struggled.

Scarborough released his neck, struck a blow across the

man's face that hurt his hand, then pulled the man back off the sidewalk and into the garage driveway. The garage wasn't residential, and it was dark inside; nobody would hear this fight unless they happened along by foot or car.

And Scarborough intended to make this as short as he could. He wasn't in bad shape for a fifty-year-old, but he knew he couldn't keep this up for long.

"Tell me, asshole, or I'm going to *kill* you!"

"Get off!" said the man, curiously placid for a guy who was getting beaten up. "This is not advisable!"

"What? Not advisable! What the hell is going on? Who are you?"

The man said nothing.

Scarborough, anger driving all fear from him, struck the man again across the face. The man's head whacked back against the concrete, and he went limp. Unconscious. *Damn!* Scarborough hadn't intended to put him out!

Still . . .

Scarborough felt around for the man's back pocket and the bulge of his wallet. Yes, there it was. He pulled it out. Black leather. The usual man's wallet. Scarborough opened it, expecting some kind of badge or ID.

Nothing but credit cards and a driver's license.

He pulled the stuff under a sodium light.

The driver's license was from New Mexico.

The name on it was Philip Roscoe Newton.

There was an address as well, but Scarborough didn't get the chance to read it. A blow came out of nowhere, landing on the back of his head with a sickening thud that drained all light from his eyes, and all awareness from his mind.

When he woke up, Scarborough was lying in the garage doorway.

The first thing he thought was that he'd drunk too much and had an accident. He felt sick and disoriented. Scarborough had woken up hung over before, but he'd never been hit hard over the head, so he had nothing to relate it to but hangovers.

His head hurt badly, and he felt dizzy, though, and so it only took a few seconds before he realized that this headache wasn't from drinking. The memory flooded back . . . The man . . . The struggle . . . The wallet . . .

The wallet. Scarborough looked down at his hands.

No wallet there, but clasped in his palms was a piece of paper.

Groaning, Scarborough got to his feet, feeling at the back of his head. No bump. And when he brought his hand back and looked at it, there was no sign of blood either. He shook his head and took a deep breath and then staggered over into the pool of light.

He opened the paper that he'd been clutching in his hand.

Its block print read: "The time is not right. Be careful, Scarborough. Be *very* careful."

Scarborough turned it over. That was all it said; there was nothing at all on the back.

His headache was fading to a dull throb, and he felt as though he could walk now.

A deep free-floating anxiety gripped him.

Who *were* these people?

What did they want with him?

He crumpled the paper, stuck it into a pocket of his jacket, and started to slowly make his way back to the car.

He could almost feel their eyes burning through his back.

Who were they?

Somewhere, deep inside, he had the feeling that if he could find that out, then a lot of the byzantine puzzle that faced him would fall, at last, into place.

Somewhere in the harbor, a boat's whistle sounded. Above, a cloud folded around the moon like a fist.

Scarborough walked back to Fells Point, the taste of blood in his mouth, and determination bellowing in his veins.

CHAPTER SEVENTEEN

It was dark and quiet on Wright-Patterson Air Force Base as Lieutenant Marsha Manning stepped out from her office at nine-thirty in the evening and made her way to Storage Area Number 34, her magnetic ID-card gripped in her hand.

The sky was partly cloudy, but fortunately with no taste of rain in the air. It had gotten slightly chilly, though, so Manning wore a sweater underneath her parka. The smell of jet exhaust and the grease of the motor pool touched the air, a gritty edge to the otherwise sweet smell from the beds of new-blooming lilies besides the Officers' Quarters.

No comment had been made about her working late—her fellow officers knew that she had a deadline, and besides, she worked late often. So now, with the base practically deserted, she felt safe stepping out to check up on this bin containing the classified material.

It had taken her most of last night to break into the ID department and to cobble up the necessary programming to give her card access to the lock, and also to make sure that using it wouldn't ring a red light somewhere.

Fortunately, this particular classified material wasn't of an extremely high level. If it had been, it probably wouldn't be on Wright-Patterson, anyway, and there'd be a human guard out there. The programming was primitive and relatively simple, but Marsha was careful and took her time. She found the proper code and then gave her own card access, writing in a temporary subroutine that would conceal her entrance—a subroutine that she could easily call up again and erase, so that there would be no computer evidence of tampering, either.

She felt excited. She felt as though she were walking on eggshells. She was worried, too, about Everett Scarborough. There had been no calls from him, although she had gotten a call from Camden, who'd filled her in on Scarborough's activities and troubles. Imagine, his own editor, a traitor. One of them! She felt happy that her own loyalty was intact; Scarborough was a good man, a man who stood for something honest and forthright, and there were precious few people these days who could measure up to *that* ideal. Maybe that was one of the things that attracted her to him, that sense of *righteousness* about him.

Camden had agreed to keep in touch. She'd decided not to tell him about her little mission tonight. She trusted Scarborough, yes, but she didn't really trust Camden. Who knew, her antics might pop up on the headlines of the *Intruder* in a couple of weeks—"Secret Informant Reveals Quest for Crashed Saucer Spacemen!" It wouldn't take much for the brass to figure out who that informant was, and then wouldn't her ass be in a crack!

No. Best to keep this to herself for the time being. If she found anything interesting, she'd tell Scarborough. If she didn't, she probably wouldn't even mention any of it to him. He had other things to worry about.

She'd walked most the way there, when suddenly the twin headlights of a vehicle roamed around a corner, and the familiar engine noise of a jeep growled her way. She just ignored it, eyes kept straight ahead.

The jeep stopped.

"Hey! Marsha? That you?"

Manning felt an involuntary contraction of her muscles, as though she'd been caught in the act. But no, that was silly. No way whoever this was would know where she was headed or what she was doing.

She stopped and looked into the jeep. Yes, the voice was familiar. "I'm sorry..."

"Geez, I guess even good looks don't shine in the dark. It's Pete, Marsh. Pete Adams. you know, the guy that bought you a couple of nice dinners, took you to a concert... You forget so quickly. Say, I'm off duty in ten. What about a drink at the Officers' Club?"

Lieutenant Peter Adams. One of the several officers she'd

dated since she'd been transferred here to Wright-Patterson. A nice, bland-enough sort, she guessed, and though she'd had a couple of nice kisses with him, Marsha made it a policy to not get sexually involved with an Air Force guy unless she simply couldn't stop herself. And with Pete, that was easy.

"No thanks, Pete."

"Rats. Guess a poor bachelor's guy's going to have go home alone to his *Batman* comics. Say, what are doing out here in the boonies, anyway? Need a ride or something?"

"No. Just taking a walk. Anything wrong with that?"

"Hey, don't get defensive! I just asked. Sheesh . . . Tell you what, I'm going to stop off at the club anyway, see what's goin' on. If you want, drop by and I'll buy you a Singapore Sling. The kind with the little umbrellas, you know?"

"Thanks, Pete. I'll think about it."

"*De nada. Adiós*, Marsh. Watch out for Air Force rats down there in the storage areas. They're big suckers."

Marsha cringed as the jeep roared off into the night.

Rats. Damn. She hated rodents of any kind, and rats were number one on her hate-list. And now she had to go rooting around in some dark and dusty storage space.

Right. Thanks, Pete, she thought. Just what the girl needs to send her on her way. Nonetheless, she struck out again for her destination.

The old hangar was hunkered down in the darkness like a sleeping giant. Its corrugated metal sides glimmered dully beneath the moon and the few stars that broke through the haze unclear otherwise drifting in the sky. The area smelled of wet tarmac and old tires, of rusty fences and the forgotten past.

Manning walked the last few paces swiftly.

There it was. The gate. The card box.

She turned on the small light provided.

She pulled up her ID card and stuck it in the slot, and fingered out her code.

The gate clicked.

It was open now. Manning breathed a sigh of relief. All she had to do was to push it open, go in. She turned off the little light, braced herself, and took a step toward the gate.

Headlights fixed her as the roar of an engine approached.

Manning's heart jumped to her throat.

Damn! Was this some kind of MP patrol in this area that she hadn't known about?

She pulled her hand away from the gate and swivelled around to confront the new arrival.

"Yo! Marsha! Sorry to bother you, but I didn't think you'd be coming to the club and I just remembered. I got two tickets to the theater next weekend and no date! Whatcha say?"

Manning relaxed somewhat. Pete. Only Pete Adams.

"What's playing?" she said, sounding cool.

"It's a revival of some Noél Coward play. All Brit actors, touring company. Supposed to be real good."

She let him dangle for a moment.

"When again?"

"Saturday week."

"Okay, Pete. I'll go. Thanks."

"Great. Say, whatcha doing here?"

"Just curious. Making a spot-check of security arrangements for storage. Part of this new assignment I'm on."

"Oh . . . so *that*'s why you're in the area. It didn't really make much sense, but Pete Adams's happy voice showed he was so pleased at the acceptance to his offer, he wasn't thinking about much else. "Say, what about a late dinner at my place after the show, too. Some vino, a little light jazz . . ."

"That would be very nice, Pete. I'll look forward to it."

"Great. I'll call you this weekend with the details."

"Fine. See you, Pete."

The engine hummed to life again, and wheels screeched away into the darkness.

Manning allowed herself a heavy sigh of relief, and then took another few deep breaths for good measure. She walked to the gate, opened it. It squeaked loudly from rusty hinges, but there was no one in the area to hear. Still, she moved it slowly and carefully, and then placed it back much the way she'd found it, only not locked.

She stepped lightly over to the side door of the hangar. The front access doors, through which planes had been dollied in and out years and years ago, had been cinderblocked-and-cemented shut.

The side door of the hangar was locked as well. However, it too had a computer access box, albeit with a separate entry code. Manning had predetermined that code and she carefully tapped it in now.

A whir, a click.

The locking equipment of course was much brighter than the old door that surrounded it; it was much newer. It opened easily at Manning's touch, and she pushed her way in.

Dust.

The interior smelled of dust and age. She'd brought along a small flashlight, and clicked it on, using the cone of light to search out the main light-switch. After finding it, she first closed the door behind her, then clicked the light on. A series of dim bulbs lit up, revealing aisles of shelving, wire cages, and containment boxes. She had no worry of these lights being seen from the outside. The hangar had no windows.

It looked like an abandoned warehouse. There was no sign of recent activity here—all the stuff stored seemed to be old files, boxes, and such, along with labelled crates.

Manning smiled to herself. She was reminded of the ending of Steven Spielberg's *Raiders of the Lost Ark*, where the Ark of the Covenant is stored in a vast government warehouse.

The bins were wire cages and compartments, located on tiers of shelving that rose up to the arching roof. Each compartment was lettered. Manning knew which compartment she was looking for. She'd memorized it.

Bin 27.

As she walked, the echoes of her low, regulation shoes scuffed through the hangar, causing a subterranean effect. She checked the numbers at the ends of the shelves. Finally, she found the aisle she was looking for.

As she started down the dim, shadowed alley, she heard sounds from the opposite end.

Snicks.

Scurries.

Those rats that Pete Adams had been talking about?

An involuntary shudder hurried up through her spine. She stopped in her tracks. The sounds had disappeared.

Rats. She *hated* rats. She forced herself to continue.

The dimness in the aisle was such that Manning had to fish out her flashlight again to read the numbers on the bins. Halfway down, toward where the hangar's airplane-doors had been, the bins were much larger, with no shelving tiers on top of the floor. She found the bin marked 27.

It was open. Beside it was a label.

Manning read the label with her light:

> CLASSIFIED. ROSWELL, 1947. Materials discovered on Brazel Ranch. See: Research composition analysis reports, 1947, 1955, 1960, 1965, 1970, 1979. Analysis reports via Special Lab, White Sands, New Mexico. Remainder of materials from Roswell crash, Adjacent Bin 27.

Manning opened the cage of the bin, and flashed her light inside.

The reflection almost blinded her a moment. She had to aim the light off to the side wood wall to see properly. Sure enough, what lay on the floor was a big pile of highly reflective material. It looked at first like metal, but then she saw that whole pieces of it looked crumpled and a little dirty or burned in spots, like aluminum foil used in some massive barbecue pit. She stepped into the bin, knelt before the stuff, and picked a length of it up.

Remarkably light, despite its thickness. And strong! Not like foil at all.

She went further in, picked up another larger piece.

With a shriek, a fat, furry body hurtled out from the darkness beneath, skittering away into a corner.

Manning couldn't help the scream that burst from her. Her heart was pumping hard, and she stepped back against wall, trembling.

It took her almost a full minute to recover.

C'mon, Lieutenant, she told herself. *The rat's more afraid of you than you are of it! Get a* grip! She rallied and stepped back over to the metallic material. She would have liked to take a piece of it with her, but they seemed simply too large to haul away, and of course there was no way she could possibly hope to carry one of them back with her to the car!

It didn't make any difference. The significant thing was that this stuff didn't look like anything Marsha Manning had ever seen before. That was something she could definitely report to Scarborough. Also, there was the label. "Roswell Crash," it had said. *Something* had fallen in New Mexico, and this proved it. But the question was, what was it?

Maybe whatever it was lay in Adjacent Bin 27.

She left the pile of odd metallic stuff and went a little farther down the row, to a door with the appropriate label.

This was a very large bin—large and deep. Perhaps thirty yards by another twenty. Inside was a dark form.

Manning put her light on the form. It was perhaps eight feet high, and stretched out another twenty-five, tapering down from a hump in the middle. But the form was irregular, as though it were just a large pile of refuse . . . Looking closer, manning could see that it was covered by a huge tarpaulin.

She tried the compartment's door. It was not locked, so she went in.

The tarp looked like the sort they covered jets with, but quite old—certainly older than anything mechanics would use nowadays. She lifted a frayed edge, wary of any rats running out.

Nothing.

She pulled up the edge of the tarp further, and there appeared to be more of the strong, foil-like metallic stuff, attached to odd-shaped struts and fixtures.

Well, the tarp's light enough, she thought. Might as well go for the whole kit and kaboodle!

She pulled the covering up higher.

She had to walk along the length of the thing to uncover it sufficiently to have a proper examination. When she was finished, she backed up and directed the light of her flashlight, panning it to illuminate the object section by section.

When she was finished, she turned the light off.

She started pulling the tarp back on, lest someone realize that she'd been in here. Halfway finished, she realized that her hands were shaking. She had to stop for a minute, to get her bearings.

Dear God, she thought, wanting to get this over with, wanting to get the hell out of here. Dear God in heaven!

From somewhere in the cavernous hangar echoed the nocturnal scamperings of a rat.

CHAPTER EIGHTEEN

"Hello?"

"Ed?"

"Hey, Ev. Great. Right on time. How's it going out there in the cold?"

"I'm still alive."

"You sound good, Ev. You sound *real* good."

"And you sound like you've got something for me."

"Not a whole lot, but it's something for you to work on, I guess. But you're sure you don't want to meet somewhere? No, maybe you're right . . . maybe that's not such a good idea for right now."

"What do you mean?"

"Part and parcel of what I've got for you, Mr. Scarborough. I'll start with the bad news. Unfortunately, I have absolutely no indication that the CIA, or any agency remotely connected with the CIA, from the Contras to the Afghan Freedom Fighters, have kidnapped your daughter or anyone else associated with you."

"Okay. What else?"

"What else . . . right. This Editors and Publishers business. Nothing official here, Ev, but I've done a little digging with other friends a couple of bumps up from me, and apparently Editors and Publishers is some sort of loose jargon involving the super-grades' relationship to big moneymen."

"Moneymen? I don't get you?"

"Oh, you know—the people who are really supposed to be running this country. The billionaires, the old boy networks. There's a real tie between old money and the CIA, pal—look at Georgie Bush! I think that's what those terms refer to."

"What's that got to do with the government UFO cover-up?"

"Say, I've only had a day, and I had some debriefing to do from my trip. Give me a break."

"Sorry, Ed. What else did you come up with?"

"This is where I hit some pay dirt. I did a little research on your Doctor Julia Cunningham. She indeed works for us. Neuro-chemical control research. Real science-fiction stuff. If she'd worked for the Company back in the late fifties, early sixties, she'd have been the one passing out the hits of acid. Get my drift?"

"Yes. Yes, I do."

"I still don't see the UFO connection, but she's definitely a consultant."

"You mean she's not full-time?"

"No. Apparently she's got some other kind of research and experimental interests with some universities, some connections with the NIH and the NIMH."

"Yes. That's where I met her. The National Institute of Health. Hurry up, Ed. I don't want to stay on much longer. I know this is a clean line, but allow me my paranoia."

"I understand. According to records, Dr. Cunningham is on assignment now at a government research base in New Mexico. Actually, it's on Kirtland Air Force Base. Heavy-duty psycho-control research rep, dating back from the forties. And the CIA have been using it ever since Truman whacked our little fanny."

"Thanks, Ed. Can you be at the phone this time tomorrow night?"

"You bet. You know, that offer of man-to-man help still stands. I got some nice leave time coming up."

"Thanks, Ed. I may have to take you up on it. Bye."

Scarborough hung the pay-phone up. He picked the change out of the box and started feeding the thing again, along with more quarters from the stack he'd gotten at the bank in Fredrick, Maryland, today.

Scarborough was in a bar in Uniontown, Pennsylvania, now. He'd just had a light draft beer and steak and a baked potato. He'd been trying to call Marsha Manning since he got here, but the ring on the other end had vibrated into nothingness.

He figured he'd try again.

He dialed out her number.

Scarborough didn't just need to talk to her, he needed to hear her voice. Today had been pretty rough.

Who were the men following him?

All day long he fancied he saw them from the corners of his eyes. But when he turned—either it was someone else . . .

Or worse, no one else at all. A flutter of cloth. A post. A sign.

Who were they?

He'd read that note they'd crammed into his hand a hundred times.

Not time yet . . . Not time . . .

Not time for what?

His mind spun off into a hundred tangents, a hundred possibilities. Were they government officers, working to expose this cover-up themselves? Were they independent? Were they reporters, working a big story? Were they military? Were they *foreign* agents? Were they members of the Publishers themselves, playing out some complex and twisted game?

Whoever they were, they were following him. They were toying with his life. They considered him some kind of pawn in this game of theirs, and Scarborough didn't like it.

Not one bit.

Finally, he just had to give up thinking about it, and give his fevered brain a rest.

What he wanted now wasn't just answers. He wanted a moment's peace, a short time of sanctuary. The only place he felt safe right now, was in Ohio. With Marsha Manning. He had a peculiar affection for Marsha. He'd developed some kind of odd dependency he hadn't felt with a woman since . . . Well, since his wife was alive. Normally, he'd have run away from that feeling, but now he was running to it.

Times had changed.

So that was where he was headed . . . Marsha Manning's house, in Ohio.

And then . . .

And then he had an idea of where he would go next.

On the sixth ring, the phone was picked up.

"Hello?"

"Marsha. It's me."

She was out of breath. Scarborough could hear her taking

gulps of air. Finally: "You must have heard my telepathic beckonings."

He gave her the number of the booth he was at. He was running out of quarters.

In a few moments, she called back.

He said, "Look, I'm on my way. It's been kind of crazy."

"So I've heard. Camden filled me in. You okay?"

"No, but I'm still in one piece. I . . . I need to recover a bit. I thought I'd rest awhile at your place."

"I'd love to have you. But I don't think it's a good idea, Everett. I really don't."

Scarborough felt a momentary pang of rejection—and then he realized that Manning was mostly interested in his well-being.

"What happened? You see agents watching your house?"

"No, but I can't be sure any more. This is *big*, Everett. A lot bigger than I thought it could ever be, and I think that we must be very, very cautious. I've used my computer know-how to check this phone for taps, and it's clean. But after tonight . . . Well, let's just say, although I did my best to cover my tracks, they might coming looking for me, too. And if they find you here as well . . . The game is up."

"What are you talking about?"

"Did you know, Doctor, that Wright-Patterson has a UFO rep?"

He had to think about that for a second. "Oh. Sure, the crashed saucer. From Roswell, New Mexico. UFO-ology 101. Nonsense. I looked into it myself. Hearsay. Just like a lot of the UFO mythology."

"Think again, buddy."

"What?"

"I've been reading up on the subject and that little bit of information hit me square between the eyes. So, here I am, right on Wright, so to speak . . . I checked the classified inventory with my computer prowess. The files said there was something in a hangar marked Roswell, New Mexico. So I finagled clearance for my ID and I went and had a look-see."

"You're not going to tell me you found a flying saucer, are you, Marsha?" His teeth were gritted, but he tried to maintain a jokey, light attitude toward this, the attitude that had put him in such good stead all those years as a lecturer and a guest on TV shows.

"Not exactly. Let's just say I found something in that hangar that was made of stuff I've never seen before, that *looked* like a flying saucer that had made sudden impact with the ground—and had *writing* on it. Symbols I've never seen before in my life. Very *alien*-looking symbols."

Scarborough felt the by-now-familiar shifting of reality . . . the tearing of its very fabric.

"Everett? Everett, are you there?"

"Yes. Yes, Marsha, I'm here. Look, let's just set this aside for a while . . ."

"What do you mean, set it aside. This is *important*! The Air Force—the CIA . . . The whole government . . . They've been *hiding* this! The earth really *is* being visited by beings from other planets, Everett! Do you still refuse to accept that?"

"I don't know, Marsha. I didn't see what you saw. I can't make a judgment yet. That's not what I meant though, so don't get upset."

"It's difficult *not* to, with your stubborn refusal to face facts."

"Okay, let me give you a few facts that *I've* come up with, and you can make your own judgments from there."

"All right, Everett. You just really tick me off sometimes."

Scarborough had to smile to himself. "I guess the feeling's mutual there, Marsha."

He told her about the men following him, about his run-in with them last night . . . He told her about contacting his friend Ed Myers, and the information the CIA operative had been able to dig up. But more important, he told her his thoughts and conclusions about the covert and frightening operation he suspected the so-called Editors and Publishers to be involved with.

"But why would human beings want to *do* that kind of thing?" asked Marsha, after a short silence between them.

"It all makes a kind of twisted pattern. The CIA has been interested in mind control for years. As a weapon, as a tool. Oh, after the investigations, they said they stopped it—but you'd better believe under Reagan and Bush, the Company would be given the freedom to start up things again, full throttle. That farm in Iowa pretty much indicates they'd been doing this kind of thing since the beginning."

"I still don't understand."

"Disinformation, Marsha, as well as mind-control research."

"You mean, all these alien-abduction stories . . . Strieber, Schroeder . . . the ones reported in those Hopkins books . . . You're telling me that the CIA is responsible?"

"Or some outlaw branch of the CIA that maybe even the bulk of the Company doesn't know about."

"But why? What information are they covering up . . . Wait a minute . . . I'm starting to see . . . Goodness . . . Everett, do you think that they're concealing *actual* alien activity?"

"I don't have enough information to come to that conclusion. But that is a possibility. A possibility made more real by your work, Marsha. I'm sorry about my reaction. I guess it's from my many years of skepticism."

"Then you think that there are aliens on earth, Everett?"

"No. I still don't believe that at all. The facts aren't all in, Marsha, and besides—I'm much more concerned about my daughter, and the operation that kidnapped her. I shudder to think what they're doing to her. And to Tim . . ."

"Well, I guess that's about as far as I'm going to get with you for now. But I'm not going to let this business drop."

"Good for you. There may be a connection. As for me, I guess I should head on to my next stop."

"And where's that?"

"I have reason to believe I know one of the people involved in the abomination. A Doctor Julia Cunningham. Ed Myers tells me that she is indeed on the CIA payroll. And she's in New Mexico, Marsha. At a place known for this kind of research. That's where I'm headed. I've got a dreadful hunch that's where they're keeping Diane."

"New Mexico? Anywhere near Roswell?"

"Look, I don't have time to investigate crashed saucers . . . but as a matter of fact, it's on Kirtland Air Force Base."

"Well, call me tomorrow night, because I'm not through with this!"

"What do you mean?"

"Just call me tomorrow night, Everett. Okay?"

"Look, Marsha, if you're up to something, you really should let me know about it."

"Tomorrow night, Everett," she said, a little coldly.

And hung up.

One more call, and then he was going to drive aways more and find a place to sleep.

He went to the bar and got some more quarters, along with a Coke. It was just a roadside bar, with a deer's head above the cash register and jars of pickled pig's feet and pickled eggs on the counter. The steak had been tough but filling, and the potato had been a huge Idaho.

"You got some serious phone-calling to do, mister," said the bartender, a sleepy-eyed, middle-aged man who looked as though he sampled his wares frequently.

"That's right." Scarborough pushed out two quarters for the Coke, and four for a tip. "Anybody waiting to use it?"

"Nope. Sorry if I sounded critical. I like to mind my own business. Just making a comment."

"That's okay."

There were only about ten customers, but what with Dolly Parton and Buck Owens coming from the jukebox and the clicking balls and curses coming from the pool table, to say nothing of the clatter of dishes in the kitchen, Scarborough was able to get quite a bit of privacy back in the dark corner.

He picked up the Coke and sipped. Shaved ice. Good. Didn't get too many bars that gave you shaved ice, and that was the way the Scarborough liked his Cokes, when he drank them. He carried his frosty glass back to the phone, set it down by the ripped phone books, put the quarters beside the phone, and made the call.

He hit on the second ring.

" 'lo."

"Camden. Scarborough. Call me back at this number."

"Yikes. Hey, man, I'm just about to go out. Celebrate."

"Camden! We need to talk."

A momentary pause as the man, who sounded a little foggy, thought about the situation. "Okay. Gimme the number."

Camden called back. "Hey, Scarborough. Sorry. I was thinking my long distance phone company had locked me out. Guess they didn't."

"Why didn't you say that?"

"Uhmm—I figured you'd have called back if you didn't hear from me."

"You bet I would have. Camden, what are you celebrating?"

"Great trip to New York. Picked up some interesting stories to keep my readers happy—until I get to split, anyway."

"Well, good for you. I'm glad you talked to Marsha and filled her in. Thanks."

"No problem. Where are you?"

"Not important right now. I think I'm onto something, Jake."

Scarborough told the reporter much the same thing he'd told Manning. However, he didn't tell him what Manning was up to—it somehow didn't seem to dovetail sufficiently; and besides, he had to keep Camden's mind on relevant business. You toss a UFO reporter any new stuff on crashed saucers in New Mexico or the Majestic 12, or any of the chief concerns or catchphrases of the saucer world, and they salivate like Pavlov's dogs.

"Geez. Sounds hot. Sounds like I should *be* there!"

"Yes. I could use your help."

"Listen, though, I got some serious writing to do beforehand. I take it you're going to wear some tread out to get there."

"Yes."

"Okay, that gives me a few days, right?"

"I guess so."

"Good. Where you going to be in New Mexico?"

"I don't know yet."

"Look, I'll just be ready to fly out there. No problem. The Southwest is always a good place for saucer stories—mystical Carlos Casteneda stuff and all that. I go out there all the time, so nobody will say anything. I gotta get these stories done first, okay?"

"All right. Thanks, Camden."

"Sorry to hear about that little run-in with the Hardy Boys in Baltimore. And I haven't got a clue who they are, either. Nothing like it in saucer-lore. Men in Black generally just intimidate; they don't clobber you over the head."

"I think we're dealing with much more than just lore, here, Jake. We've got a serious government conspiracy that misuses and even *harms* United States citizens in the interest of God-knows-what perverted cause."

"Say, you know, I've been thinking about that, pal. You remember that guy, Harry Reynolds . . . the shortwave guy who disappeared last month. Think maybe it was these Editors and Publishers that got the old 'Klatuu of the Airwaves'?"

The notion struck Scarborough like a thunderbolt.

"Scarborough! Hey, chum, you there, or did the Grey Men get you?"

"No, no, I'm here. I was just thinking. Harry Reynolds. Yes, I remember him."

"You better. A real UFO crackpot. One of the classics. I met him. Nice enough old guy. He had this prosthetic foot. Got the real one blown off way back in Korea. Used to wave it around sometimes at UFO conventions. What a hoot. . . . Hey, you're not taking me seriously, are you?"

"Reynolds lived in Iowa, right?"

"Yeah. Dubuque. So what?"

"I'm just remembering, that's all."

"What, you're remembering reading about it? You don't want to investigate it do you—Hell, old Harry probably just crossed over to Canada with some young nurse to start a new life."

"No, I don't think I was all that aware of the disappearance until MacKenzie showed me a letter he'd gotten."

"Letter? What kind of letter."

"A letter suggesting a permanent alien abduction of Reynolds. A letter from a guy in New Mexico. Now *what* was his name? As I recall it sounded Russian. Mashkin! that's it Mashkin! You know of a guy named Mashkin in New Mexico?"

"Well, yeah—a real letter-hack. Writes to all the publications. I've used plenty of his letters in my column. Mashkin. Yeah. Walter Mashkin. From Albuquerque, New Mexico!"

"Jake, I hate to say this, but you're proving to be absolutely invaluable."

"Told you!"

"You go out and buy yourself one on me, Jake Camden. We'll keep in contact through Marsha, but don't be surprised if you hear from me this weekend with arrangements to meet you in New Mexico. I've got a real good feeling about this, Jake. I really think Diane's there in the government facility—either that, or in a secret installation not far away, one that these Editors and Publishers use for their experimental operations."

"Sheesh. Hotter and hotter. What a fucking great story!"

After hanging up, Scarborough went back to the bar and ordered a shot of Jack Daniel's whiskey. He carried it back to the pay phone, lifted it up and saluted the air.

"To you, Mac."

He drank the whiskey, and then he dialed Directory Service for New Mexico.

CHAPTER NINETEEN

When Camden saw Joseph Donohue, the resident counsel for the *Intruder*, waiting for him in Kozlowski's office, he knew there was trouble.

"Hey, Joe! *Que pasa!* How's the wife and kids and the barbecue?" He kept up the breezy attitude as he strolled along the plush carpet and plopped down confidently into the chair in front of the desk, putting a leg up on one knee and putting his hands behind his head.

There was trouble, yes. Exactly what kind of trouble, he couldn't say, but with lawyers, it was best to strike a confident pose, or they'd tear you apart with the flick of a phrase.

"Jake," said Donohue, nodded in his armchair and then turned to look at Kozlowski. "Shall I tell him, Mr. K., or do you want the honors?"

Kozlowski's cigar was dead in the ashtray, which was a bad sign. He just grunted and looked down at his nails.

Taking the signal from his employer, Donohue turned to Camden and fixed him with an authoritative look. "Jake, it's about the first part of your journalistic effort concerning Dr. Everett Scarborough."

Camden could see a marked-up copy of the eight-page first installment in front of Kozlowski. Another one was held in a Gucci-leather business folder on the desk by Donohue.

It was Monday afternoon.

"What, I accidentally slander somebody? It's good though, huh? Really good?"

"Fucking best piece I've read in a long time," Kozlowski muttered, not looking at Jake.

Camden smiled. Maybe this wasn't so bad after all. The old man was just floored, that's all. He'd showed him, and Kozlowski was just stunned that he had such a good reporter.

"Yes, it is well written, Jake," said Donohue. "However, Mr. Kozlowski asked me to examine it for potential problems, and I found a couple. Significant problems, Jake. Problems that we're going to have to discuss in some depth."

Donohue was a heavyset man in his early forties with an overly manicured look that fit as snugly around him as the three-piece grey pinstriped suits he favored. His face and hands had a well-fed sleekness to their even tan that looked healthy in a sharkish kind of way. He had a big nose and a receding hairline, which made his narrow eyes look even more deeply receded and beady than they actually were. He was the kind of lawyer, Jake Camden knew, who had tied Florida down in the last twenty years of her incredible real estate and environmental rape that John D. MacDonald wrote about so well. He'd gotten into legal trouble a few years ago and had to make a career switch, and that was how Kozlowski had gotten him. He was the kind of lawyer who cost you a bundle on either side of a consulting table or judge's bench, and Jake supposed Donohue must be good, for Kozlowski to want him full-time. Privately, though, Camden always despised the man.

"Okay, no problem. We just change a few names to protect the innocent. Maybe I got too many facts in there. Change it however way you want to, guys. It's your piece now, right?"

This was actually what he had counted on, Kozlowski rewriting the article, though he didn't expect Donohue to get in on the act. Didn't make any difference, because this way he just didn't have to write another article when he showed the *New York Times* the story.

"No, that's not what we're talking about, Jake," said

Kozlowski. "We're just not going to be able to run the story, period."

"Huh?" Jake blinked, surprised but not quite knowing what to make of this announcement.

"The really big problem, Jake," said Donohue, "is that this article reveals the fact that you have aided and abetted a criminal wanted by the FBI. As an employee of the *Intruder*, this explicit admission could possibly endanger our legal standing."

"What? That's hogwash! I'm a *reporter*! I'm on a *story*! Reporters talk to criminals all the time."

"You're withholding information which could lead to this man's arrest, Camden. This is against the law. The *Intruder cannot* condone this activity, implicity or explicitly."

Jake looked pleadingly at Kozlowski. "What's going on here? You *loved* the story, before and after you read it!"

"Sorry, Jake. I gotta go by what Joe says. He's my lawyer, and I gotta trust him on this."

Camden shook his head. "I can't believe it. You're not even going to get a second legal opinion?"

"I can show you the laws involved if you like, Jake," said Donohue stiffly.

"The wrath of Hollywood stars and crackpots we can hazard," sighed Kozlowski. "But after thinking about it some, I just can't fuck with the U.S. government. Especially the boys that get to tote the guns."

Jake suddenly realized why. Kozlowski's past—he probably had to keep his nose extremely clean. Still, actually, come to think of it, this wasn't that bad. He was confident that the *New York Times* or the *Washington Post* would print the story. And *this* way, he wouldn't have to go against Scarborough and squawk the story before things worked themselves out.

He wouldn't have to go against Scarborough's wishes, and Camden actually felt pleased at that. He liked the guy, and the last thing he wanted to do was to screw him over.

"I'm very disappointed," said Camden. "But of course I understand." He reached over and tapped the article. "Pretty hard-hitting stuff, this, and more to follow. Tiger by the tail, huh, chief. Well, that's okay, as long as the check clears."

Kozlowski scratched his nose. "It won't. We put a cancel on

it, Jake. It's a damned good story, but we can't pay for what we can't use."

Camden's jaw dropped. "What?"

"That's correct, Camden," Donohue said officiously. "I called the bank myself."

Camden stood up and poked a finger at his boss. "You *promised*!"

"I promised money for something that would up circulation. Since we're not using it, it won't up circulation. So, find us another hot story."

Camden could barely speak. He felt as though a giant had taken the sides of his chest between massive fingers and was now *squeezing*. *Oh fuck*! he thought. *Plentenos! I won't have the money for that bloodsucking Cuban! He'll kill me!*

"But boss, I have obligations to meet! You know I'm in a jam."

"Yes. But I think the next part of what we have to tell you is going to solve that, Jake. And of course, it's not like you're being fired. You'll still be on payroll."

Camden blinked. What was the old guy talking about?

"Next part? What next part? I don't understand."

"Joe, you better break it to him. You're the one who made the call."

Donohue stiffened in his chair, as though bracing himself. "Camden, I was extremely alarmed at some of the material in this article. I, for one, don't believe the story you tell. Not for a moment. And since, in reading it, I, too, became privileged with the knowledge that you are aware of the whereabouts of this sought-after killer, Everett Scarborough, I took it upon myself to call up the closest office of the Federal Bureau of Investigation. They're sending a man over later this afternoon to ask you the necessary questions."

Camden couldn't believe what he was hearing. "You *what?* You mean to tell me, you sicced the *Feds* on me, you scummy *bastard*!" He swivelled to Kozlowski. "How could you allow him to do this! They're going to haul me away!"

Kozlowski shrugged. "So you'll be safe from your drug lord or whatever. I'm doing you a favor here, Jake. You'll thank me later.

"I didn't call the Feds, Jake. That rabbit was out of the hat.

But as long as they're coming, you might as well answer their questions."

"Don't you see, I *can't* do it! I have a man *depending* on me! Didn't you read that piece? Can't you see that I've made a *commitment*? Can't you see how *important* this is?"

"C'mon Jake. I know what's important to you. Money, booze, and women. Don't try to pull the wool over my eyes, you're not really choked up over this guy. You're in it for yourself. Look, maybe you can sell your story somewhere else. It's just too much of a risk for us here. And Jake—take my word, it's really for your best interests to cooperate with the authorities.

"Well, I won't!" Camden stalked out the door.

"Jake. Where the hell do you think you're going?"

Typewriters stopped in the main editorial room. Heads swivelled.

"I'm getting the hell out of here, that's where I'm going!" Camden yelled back. He turned wild eyes on the bull room, full of editors, writers, assistants, and secretaries. "The guy's gone totally bugfuck, people. He's got a couple of *narcs* coming over here any moment, and they're gonna *comb* this joint!"

Pandemonium broke out.

As chairs fell over and drawers screeched open and people rushed to the ladies or men's room to flush their stashes, Jake Camden hurried down the hall, not even looking behind him to see if Kozlowski's red face was poking from his office door.

Fuck the *Intruder*!

Jake Camden had better fish to fry now. He was going to sell a book, a movie—and have a dynamite series in some *prestigious* paper.

Of course, he had to get the hell out of town, fast, or he'd either be in jail, or dead.

Jake Camden much preferred being free, and alive.

CHAPTER TWENTY

Road trip!

Jake Camden grinned as he pulled his car out onto Everglades Avenue, heading toward Route 75 north. He'd hurried back to his apartment, packed a quick bag, dug out his emergency cash cache of four hundred and twenty-two dollars and nineteen cents that he kept stashed in an old sweat sock in his closet, along with the one Visa card in his name that still worked, and an old portable manual typewriter, and flung it all into the back seat of his sedan.

Well, he thought. *I'm just gonna have to head back to New Mexico the hard way.*

It was getting on toward six-thirty in the evening, but thanks to Eastern Daylight Savings Time, the sky was still bright, the sun blazing like a god on a throne of clouds. What a state! thought Camden. You could get up early in the morning, drive east, and watch the sun rise up out of the Atlantic; have a nice day working or lolling about, then drive west and watch it set in the gulf. Camden really loved Florida. He felt more at home here than any other place in the United States. No state income tax, an amazing choice of biomes, beaches, and babes, and lots of people who could bullshit. Besides, here, people didn't give him crap about his beloved Hawaiian shirts.

Wistfully, he watched the palm trees whiz by. Somewhere in the distance, a scrub fire was burning. There was the smell of money in the wind. How long before *this* area got developed? Not long, surely; it wasn't far from Orlando, and Orlando was booming. When he made his wad on his book and movie, the first thing that Jake Camden intended to do was to sink a healthy portion of it in land, maybe hereabouts. So what if Florida was getting gang-raped, he decided, his cynicism beginning to return. Might as well come in for sloppy seconds!

He lit himself a Camel cigarette and settled down to enjoy the ride.

He'd made it about a mile out of town on Palmetto Road when a low-slung Buick land-yacht suddenly hauled out onto the road from nowhere and stopped dead in front of him. Instinct caused him to stomp on the brakes, stopping just yards from hitting the car. If he'd had time to think about it, he would have gunned it, swerved, gone on the shoulder, and made a run for it.

Three men got of the car, and Camden's heart sank to sea level.

It was a comparatively untraveled area he was in now; stretches of fields and trees being the only witnesses to what was going to happen. Jake rammed the gear of his Escort into reverse, but then got a look at the bore of the automatic .45 that Johnny Plntnos was aiming at him. He lost his nerve and stopped long enough for one of the drug dealer's henchmen to run around and pull him out of the car.

There was still hope. Not much, but still hope.

"Johnny! What the hell's the matter with you?" he said, injecting his voice with a righteous anger. "It's Monday, right? I'm not supposed to see you till Wednesday afternoon."

Plntnos holstered the .45, which made Camden feel a little better. "I got a tip you were leaving town awfully fast, Jake. Didn't sound like you were coming back, either. I don't like that, Jake. Looks as though it was true, too, yes?"

Plntnos wasn't using his Cuban accent, and he didn't look like he was playing the Spic macho-villain, either. In fact, he honestly looked upset, and not a little angry, in a quiet, unsettling kind of way. He looked like a man whose honor had been spat upon, but grieved about what he was about to do.

"Hey, Johnny, I won't lie to you. Yeah, I'm getting out of town, but it's not because of you. I've got Feds on my tail. FBI. Look I don't want to go into it—I was really going to get you the money. Just give me a little more time. And for God's sake, you'd better clear out because they just might be along any minute!"

"Feds?" Johnny Plntnos looked over to his two men, Miquel and Paco.

Miquel and Paco shrugged and shook their heads.

"I don't think so, Jake. I think it's your usual bullshit. Look, Jake, I'm tired of all this. I think I'm going to have to end it right now. You saved some time for us by coming out here onto Palmetto. Nice and secluded. But we'll make it fast, so it won't hurt too much."

"Jesus!" Camden tried to pull away from the man holding him. Muscles clenched, and the grip hardened. Not only did Camden not pull free, but he got a punch in his stomach from the other man for his trouble.

He went to his knees, groaning, the pain waving over him sickeningly.

"Too bad. Maybe it has to be painful. We take a walk now, Jake. C'mon, I see a nice path up ahead. Miquel, you get these cars off the road. Paco, you come with us."

"Christ, Johnny! Look, I've got five hundred dollars. It's yours. Christ, *you gave me till Wednesday!*"

"I give you my trust, Camden. You shit on it. Now the fifteen thousand dollars ain't important. Paco and Miquel, they think I'm a pussy for not killing you last week. I think maybe they were right. But now I make amends. People will know now. They can't fuck with Johnny Plentenos and live. Now, I fix my name, my rep. Now, I fix *you*! C'mon, let's go. The more shit you give us, the more we hurt you before we put you in the ground, Jake!"

Shit! What was going on here? Jesus, that bastard Donohue—he must have been the one to call Johnny down on him. Which meant there were contacts . . . Which meant . . .

Christ, wasn't *anyone* straight in this world?

"It was that creep lawyer, Donohue, wasn't it?" Camden spat. "What, whose pocket is in whose there, Johnny. You want to preserve your honor, I'll tell you right now, you don't want to fool with *that* snake!"

"Shut the fuck up, will you!" Plentenos's eyes blazed. "You know, I'm doing the world a big favor, Camden. I'm closing your fucking mouth, for good!"

Plentenos kicked Camden in the side. Hard. So hard that it knocked him from Paco's grasp. Camden rolled away a few yards, but the blow had been so painful that he wasn't about to get up and try and make a run for it. It was all he could do to hold onto his consciousness. If that went, so went his hope, and he'd wake up dead and buried.

Plentenos was about ready to kick the fallen Camden again when one of the henchmen yelled something, distracting him.

"What?"

"A car! A car is coming."

"Shit. Miquel, get the Buick out of the way. They'll pass."

The car, however, did not pass.

It rammed directly into the Buick, from the other side.

"Shit!" cried Plentenos. "What the *fuck*?"

The Buick swung around and gave Camden's Escort a good whack on the bumper. Camden was too low to see much, and the pain was distorting his vision, but he could make out the edge of what seemed to be a large black car. He heard the sound of doors opening, shoes clicking onto pavement. He tilted his head and read astonishment on the Latin faces canted against the sky and the palm trees.

"*Madre de Dios*!" cried the man called Paco. He reached into his yellow poplin jacket and pulled out his gun, a heartbeat before his companions started to reach for their weapons.

An explosion, like a hundred caps getting hit by a stone . . . A hole opened dead center in Paco's chest, and a gout of blood licked out like a shapeless dissolving tongue. The big man was knocked back like a broken punching bag, his knees crumpling and eyes turning up toward the sky. He slapped onto the ground right by Camden. The reporter heard the man's death-rattle squeeze from his throat.

Camden whimpered, and, half-paralyzed by pain, started crawling for the bushes for the side of the road. If this was some sort of drug vendetta, he didn't care to be among the vendettees!

Plentenos and Miquel had their own guns out now, and managed to get off a couple of shaky rounds, but as Camden had always imagined, but never wished to explore, the Latins were not particularly good shots.

A volley of bullets pocked across them like invisible stones. Miquel was hit in the abdomen and chest. Johnny Plentenos was clipped in the shoulder, and then a bullet gouged out an unhealthy portion of his forehead.

They both went down spinning like dancing lawn-sprinklers of blood.

Silence.

Camden stopped whimpering, but he didn't stop crawling toward the bushes, pain or no pain.

Then he heard the click of shoe soles against the road as the killers stepped around the car toward him.

The first thing he saw were the guns. Big, long, black things, still gripped in slender hands. There were two of them, stepping forward, weapons ready for any necessary *coup de grâce* on their victims. They both wore jeans, black shirts, black linen jackets, and sunglasses. They looked as normal as a couple of guys just stepping out of some suburban mall in Tampa. They also looked white and waspy.

Camden figured they couldn't miss him, so he stopped crawling and said, "I'm not in their gang, I swear it. They were going to kill me."

The two men, seemingly satisfied that the Latins were dead, stepped forward toward Camden, stopping several yards short. "Yes, we know, Mr. Camden."

Camden did the best double take he could manage, considering the way his body felt. "You guys know me! Who are you?"

"We suggest you get in your car and keep going, Jake Camden. We do not like this kind of nasty business, and we certainly don't care to have to kill the government agents on your trail."

Jake struggled up, astonished, but not about to turn the suggestion down. He wobbled off toward his car, but then turned.

The two men were walking back toward their car, and Jake could see that it was an early model Cadillac, and that there was another man, sitting behind the steering wheel.

"Jesus Christ. You're from the group that's following Everett Scarborough, aren't you?"

The men kept on walking silently away.

One got into the Buick and moved it off to the side of the road. The other went and opened the door of the Cadillac.

Camden's pain disappeared under a deluge of excitement. He hobbled toward the car, waving his one good arm and yelling "Hey! Wait! I want to *talk* you! I can *help* you, for God's sake Just tell me who you are? What's going on? What *is* this strange UFO conspiracy that's going on? Gimme a break, guys Let me show you what I can do for you."

All the time he was thinking, *Christ, layer upon layer, wheel within wheel—I'm sitting not on the story of the decade, it's the story of the* century!

The standing man turned toward Camden and took off his sunglasses.

His eyes were like ice—deathly cold with determination. Camden had to stop in his tracks, that stare was so effective.

The man strode forward and met Camden halfway. The journalist could see that though the man wasn't any taller than he was, that jacket was hiding some pretty intense muscle, compact and wiry. Then the man grabbed Camden's shirt and pulled him toward his face with his left hand. The right hand still held the gun, and the man lifted this and inserted the gun into Jake Camden's mouth. Metal rattled against teeth, and a jolt of pure fear ran through Camden like high-wattage electricity.

Camden had a sudden flashback to Clint Eastwood as Dirty Harry. He half-expected the next words to be a gritty and rough "Are you feeling lucky, *punk*?"

Instead, the man's words were soft and well enunciated.

"Mr. Camden. It truly grieves us to have to participate in this manner. Your present welfare is important. However, be advised, that if anyone is going to kill you, it's going to be us. It would be very simple to pull this trigger and leave you along with these other bodies. Now, do you want us to do that?"

Camden shook his head.

"I didn't think so. Now get in your car and drive away from here and be about your business. *Immediately.* No more questions, no more thoughts. Understood?"

Camden nodded.

The man withdrew the gun from between Camden's jaws and walked back to his car. The guy who had driven the smashed Buick to the side of the road was already getting into the back seat. The man gave one more significant glance to Camden, put his sunglasses back on, and got back into the front seat.

They waited to make sure Camden was leaving.

A shiver raced up Camden's back. He turned and ran back to his car, noting that his Escort bumper had only been dented slightly. He got in, turned the ignition key, and swung around the Cadillac and accelerated along Everglades Avenue, away from the fallen bodies of the men who had almost killed him, away from the men who had saved his life, but had threatened him with death.

It wasn't until he hit Route 75 north that he permitted himself the luxury of thought. Or perhaps that was just when the all that adrenaline pumped into his system finally started levelling off.

Who *were* those guys?

What the *fuck* was happening?

Camden noted that dusk was starting to settle down on the highway. When he reached over to turn on his lights, he realized that his hand was shaking.

He had to take three gulps from the flask of whiskey in the tape compartment before he got his hands under control.

CHAPTER TWENTY-ONE

Dr. Julia Cunningham was not by nature a nervous person. At least, not on the exterior. However, it had not always been so. The young doctor who had gone to work at NIMH those years ago had been much more fragile, a smoker who bit her fingernails and had the peculiar tendency to fidget in classrooms; the product, she self-diagnosed later, of an attention-deficit in her system. If she were a child, perhaps she would have been given Ritalin. But now, of course, there were other medications.

Doctor, heal thyself.

When Dr. Julia Cunningham woke up, she was bathed in a cold sweat. The sheets of her bed were snarled and knotted around her, and her teeth were clenched. Her heart was beating wildly and she felt a high level of antiadenosine in her system. Of course, it took a full minute of wild terror and anxiety for her neocortex to reason out the situation and begin to get hold of the situation. Until then, the nervous disorder she had worked so hard to quash flared up full tilt, and waves of the nightmare's aftershocks rocked her to near paralytic frenzy.

Julia Cunningham awoke with the taste of blood in her mouth, the bile of uncontrolled fear. It didn't help much that the new quarters they'd given her were so spartan. Worse almost than a Holiday Inn. At least hotels and motels tended to have some comforting colors. This room was little more than part of a barracks, and Cunningham hated it. Back home in D.C., her own apartment was an interior decorator's dream of streamlined furniture, tasteful Marc Chagall prints, along with the subtle roll of modern sculpture. Cool, perhaps, as visitors told her. Definitely *controlled* (she had a maid come in and clean twice a week). But ultimately elegant, soft, and feminine; ultimately comfortable. The subdued tones of the rug, the paint on the walls, had all been scientifically calibrated to slip her a visual Valium every time she looked at them. No such luck here.

What she slept on was little more than a cot. A desk with a cheap military-issue lamp. A rough dresser, a couple of chairs.

She'd spent a lot of time down here a couple of years ago, but they'd given her apartment to someone else. Now, it was start-from-scratch time. If she had a day off she would have flown up to Sante Fe and bought some nice Navaho Indian rugs and artwork to soften things a bit. But she didn't have a day off. Not with the way things were working out.

Julia Cunningham lay in bed, getting her breathing back in check for a few long minutes.

The dream of her father. . .

She'd had the beating dream again.

She shivered. She thought she had purged herself of that nightmare. She hadn't had it in years, so she had just assumed she had completely eradicated the neural network which contained that dream, that memory.

But of course, as a scientist, she knew that the human mind, and human memory, did not work like that.

Still, she had tried. And would keep on trying.

Chemistry, she said to herself, like a mantra, over and over again. *It's all chemistry. Neuro-transmitters amidst the synpases, axons, and dendrites; recepting, blocking. Molecular processes—all explainable, all controllable, theoretically.*

Then why hadn't she destroyed her father?

Eventually, she was able to stagger from her bed. She wore comfortable cotton pajamas, and realized as she switched on the too-bright light in the bathroom that she was sopping with sweat. She went to the sink and stared into the mirror above it, feeling a thousand years old. The face that stared back at her wasn't old, but it was getting there. Without her makeup, she could see the crow's-feet wrinkles setting in about her eyes, the frown lines on her forehead, around her mouth. Chemical processes, surgery—all could deal easily enough with these signs of age. And of course makeup . . . thank the cosmetic alchemists for makeup! She just wished she was as effective with her neuro-chemicals . . .

She clamped down on the thought immediately.

No, she couldn't afford self-doubt. The dream memory of the child abuse her father had visited upon her was an aberration. There was a bit of root still buried deep in her subconscious, and it had grown back again. However, what had been beaten back by medication could be beaten back again—perhaps even *destroyed* this time. The experiments she had tried upon herself before had sometimes failed, driving her into the deepest depressions, but also to the highest highs. She was confident that she could deal with the problem, quite easily.

She just had to get through the next few minutes of total unmitigated and stripped-bare *hell*.

On the commode top was a large leather bag. With slightly shaking hands, she grasped it and pulled the zinc zipper. Under the flap was a collection of small cc bottles, hypodermics, and prescription boxes filled with pills from her own personal mortar and pestle. She didn't think a hypo was necessary now—that speedy treatment was only for true emergencies. No, just the right combination of the pills, ingested over the period of the next half hour, would do fine. According to the digital numbers on her alarm clock, it was only 5:16 in the morning. She didn't have her appointment until 10:00, although they expected to see her in the office at 8:30.

To hell with them. She deserved a good sleep-in.

That was, *if* she could get it.

She didn't think much about it; she really didn't have much choice. She needed the chemicals now, or she wouldn't be able to make it through the day, much less her life.

Carefully, she selected the necessary number of each tablet from their individual boxes. Even with a shaky sense of panic hanging about her, Julia Cunningham was able to cope, mentally calibrating the exact dosage.

Control.

Control was all.

Above the chemicals she used, above everything else, there existed her *intelligence*, her *will*. This was the important thing, this was the *paramount* aspect of the situation.

She was no sad drug-addict, shaking out pills or grabbing for her needle in her bathroom in the middle of the night. No, of course not, she was a *scientist*, measuring out the dosage of enkephalins, serotonin, amino acids, and sundry other chemicals, some of which she had synthesized, designed herself, to create the necessary and comfortable parameters in which her intellect could properly operate. This was the ultimate triumph of science and research, that the *mind* could physically compensate for central-nervous-system deficiencies *consciously*. All human beings had natural opiates in their biochemistry—but heroin addicts and alcoholics were like people operating on themselves with axes and hammers. Not Doctor Julia Cunningham, no. She used only the finest and sharpest of scalpels, the thinnest of laser beams . . .

She turned on the faucet and watched water pour out a few moments before she filled her glass. She took the pills and caplets and tablets one by one, taking a small swallow of liquid after each. The water was brackish, and the pills tasted faintly bitter, but their annoyance was as nothing compared to the sandpaper scratching at her nerves.

When she was finished, she carefully placed the bottles back in the purple felt, and zippered the case closed. Then, she went back to her spartan bed, crawled under the covers, and assumed a warm, fetal position.

As peace slowly salved her ragged soul, and sleep closed in like tender waves of gauze, Julia Cunningham thought, *He is coming today. My John the Baptist comes for instructions from his Master.*

And then she fell asleep and dreamed of a sea of faces, lifted in adoration toward her.

CHAPTER TWENTY-TWO

New Mexico is the fifth-largest state of the Union, a mostly rectangular, immense slab of land with Colorado and Mexico to the north and south, Texas and Arizona to the east and west. In many ways it is perhaps the most beautiful of the states—and the strangest. Everett Scarborough thought so. He'd been to New Mexico on at least five occasions, three for lectures and personal appearances on television shows, two for UFO investigative-purposes. Now, as he drove along Route 40 toward Albuquerque, he admired the mesas and the desert, the vast cerulean skies, the lonely stretches of gold and brown, along which rolled tumbleweeds and grew yucca, the state flower, and pinyon, the nut pine. In between these were occasional majestic vistas of mountains, some still capped with winter's snow. And Scarborough again felt that certain *something* he had noticed on his previous visits. That certain taste and smell in the air, that certain *vibration* of out-of-sync *otherness*. He'd tried to ignore it before, but now that he seemed to be coming more attuned to instinct and other, even less-familiar, feelings, the beautiful panoramas stretched out before him were full of an almost *mystical* sense.

Something important would happen here, Scarborough thought. This indeed was where events would climax.

This was where his daughter was!

The rational part of him was now but a tool. He had turned into a being of pure desire, hellbent upon his quest for his daughter and the truth.

The Ford was running well, as though its engine was in sync

with him, purring under the ministrations of his will. He traveled the main highways. He was in a hurry. He would have taken a plane, but the small rational part of him warned him off. Plane passengers were too easy to monitor. So he made the trip back out west in his car, quickly, only stopping for short stretches of rest.

His destination was Albuquerque.

That wasn't where Diane Scarborough was. That was where Walter Mashkin was.

Scarborough stopped at a Shell station just outside of the city and bought a map of the local area. The sun was bright and getting hot as it reached toward noon. Scarborough bought a cold Coca-Cola Classic and studied the map. After he found the road he was looking for, he filled his tank with gasoline and rode on.

He'd always found Albuquerque a pleasant town, but in the last decade it had grown so much that it had assumed much of the flavor of a generic American city—albeit without losing its roots. Scarborough passed through the Old Town Plaza, the oldest part of the city, dating back to 1706, when the place was founded by à Spanish farming community. Scarborough passed the National Atomic Museum, which he had seen on a previous, more relaxed visit, as well as the Pueblo Culture Center. Two miles farther, he made a turn onto a quiet boulevard lined with adobe style houses, and there found Number 1923, Walter Mashkin's house.

In the driveway was a battered blue Dodge pick-up truck, labeled Mashkin's Plumbing. Have Wrench, Will Travel! The back was filled with pipes, old valves, and odds-and-ends of equipment. From the garage issued a clanking sound. Scarborough shaded his eyes and made out the form of a man bent over a workbench, who was hammering on something.

He cleared his throat.

"Mr. Mashkin?"

The man stopped pounding. Scarborough could hear the tools being dropped onto wood, saw the form approaching.

"Hey, now. Would that be who I think it is?" said a lanky homespun drawl, reminiscent of Tennessee, but cut with several other accents.

"Yes, Mr. Mashkin, it's me. Can we go inside? I don't like to be out in the open."

"Reckon not, reckon not," said Mashkin. "Welcome, though. Good to meet you."

The man stepped out into the sun, offering his hand. He was a sallow-faced fellow who wore oil-stained coveralls and scuffed brown work boots. He had a beard with no mustache, which made the friendly, blue-eyed face beneath his unruly thatch of straw-colored hair even more open. Scarborough judged him to be a man in his middle fifties.

"Good to meet you, too, Mr. Mashkin. I really appreciate your help and look forward to talking to you."

Mashkin nodded, giving Scarborough an appraising look. Then, he gave a glance to either side, as though making sure they weren't being watched, and ushered Scarborough into the house.

The inside of the house was cool and dark, and smelled faintly of old beer bottles and dirty socks. Scarborough recognized the effluvium—it was a bachelor's abode, bereft of a housekeeper. Still, it was a pleasant smell, rather reminiscent of Mac MacKenzie's place after his wife left him and before he hired himself a maid, and it made him comfortable.

"You look like a tired man, Mr. Scarborough. Can I get you a beer? After we jaw a bit, you can take yourself a nap in the guest room."

"Yes. I'd like a beer. Thanks, Mr. Mashkin."

"Call me Walter. Or Walt, if you like. Just for goddamn's sake don't call me Wally. I *hate* that name."

Walter Mashkin opened the refrigerator door, revealing two shelves filled with nothing but long-necked bottles of Budweiser beer, a carton of eggs, some milk, some bread. Mashkin took out two bottles of beer and closed the door. He gestured for Scarborough to sit in one of the kitchenette chairs. Gladly, Scarborough sat and watched as Mashkin took a church-key out of the drawer and opened the beer bottles carefully, almost reverently.

"Never could take to those new-fangled twist-off higgagoodgies," said Mashkin affably as he tossed the bottlecaps into an open trash can and handed Scarborough one of the beers. "I like my beer cold and I like my beer old—fashioned, that is.'' Mashkin winked at Scarborough. Then, holding a pinkie out as though he were a cockney sipping teas, he drank a long swallow of Budweiser. Realizing that his own throat was dry, Scarborough did the same. It was almost as though the two men were involved in some ancient ceremony.

''Ahhhhh!'' said Mashkin putting the beer down on the table. ''First bottle of the day's always best. Course, I ain't no alky, Mr. Scarborough. I take Tuesdays, Thursdays, and sometimes Sundays off from my beer. ''Now then, first off, I have to tell you I'm real glad you made it here, all right.''

''Thank you, Walter. It was a pretty intense journey.''

''I'm real puzzled, though, I must say. I know your books and I know your reputation. I still don't understand how come you chose a saucer feller like yours-truly to come to for help. Not that I mind giving you help. Shit, no, I'm delighted! Mind you, I'm still curious.''

Scarborough nodded and sipped his beer. ''I guess the last month has changed me somewhat, Walter. First, however, as I said on the phone, you are close to the place where I need to go.''

''Oh, sure. But Mr. Scarborough—'' The man grinned, revealing a gap in yellowed teeth. ''I'm an UFOol, right?''

Scarborough cringed a little. ''Touché. I suppose I deserved that. Mr. Mashkin—Walter. I still do not think that we are being visited by intelligent life from another planet. However, I'm almost certain at this time that many of the UFO sightings, in this country at least, were perpetrated by an outlaw subsidiary of the government.''

Mashkin laughed and slapped his knee. ''Well, golly, and my friends down at Tumbleweed Bar and Grill say *I'm* paranoid!''

Scarborough stiffened and managed a slight smile. ''I'm afraid, Walter, that even paranoids have people after them.''

Walter Mashkin roared with laughter. ''Very good, my friend. Okay. Ice is officially broken.''

''In my case, it's the CIA. This is why I called you. I remembered your letter to Captain Eric MacKenzie concerning your friend Harry Reynolds. I think that there is indeed dirty work involved in that case.''

''What you think, that the CIA kidnapped Harry?''

''Or a part of the CIA.''

"Hmmm. Interesting. It kind of appeals to me. Not the fact that Harry's gone, mind you. The theory. But I really miss Harry's broadcasts and his letters. Met him a few times at UFO conferences. What a character! Guess maybe part of the reason this stuff interests me so much is the kind of people you meet."

"Yes. Well, it is there that I have changed my opinions, Walter. I had always scorned those who believed in flying saucers. But I believe that I can trust you because the one thing that I find in many UFO afficionados is *sincerity* and a *search* for something. I guess I've had my priorities wrong for quite a few years."

"Still and all, Mr. Scarborough . . ."

"Everett. Or Ev, if you like."

"Ev. What kind of paranoid thing is it to go to a *stranger* for help?"

"It isn't only me who needs help, Walter. As you may recall, it was *you* who asked for my friend's help in discovering what happened to Harry Reynolds. I'm offering you the opportunity to perhaps do just that. And help me along the way, in the process."

Mashkin nodded soberly. "Yeah, that sure makes sense. Sorry, but I just had to ask you straight out, Ev. Oh, and by the way, I'm real sorry to hear about the death of your friend Captain Mackenzie. Right off, I'd say I don't think you done it."

"I didn't. I could prove it in a court of law, but I don't have time to take that recourse. You see, Walter, as I said on the phone, my daughter—and her fiancé in the bargain—have been kidnapped. By this branch of the CIA, I think. And I have reasons to believe that they are being kept in an experimental installation quite close to Kirtland Air Force Base, just outside of Albuquerque."

"How come you think that?"

Scarborough told him briefly.

"Basically, Walter, I believe that this is what may have happened to Harry Reynolds. I believe that this group kidnapped him—perhaps to concoct some sort of pseudo-Encounter of the Fourth Kind, so that Reynolds could broadcast that story . . . But then, suppose something went wrong? Why, they'd have to kill him, wouldn't they?"

Mashkin's eyes were, by now, quite wide. "Wow-wee. Ain't that something?"

"Can you see that all this leads to my inevitable conclusion, Walter? All the rest, like *why* they're doing this, is in the area of speculation. But you see, *if* it's true, and if you really want to find out what happened to your friend, you can do so by helping me. Even just letting me stay here would be a great help."

Mashkin's face broke out into a large grin. "Ah, shit, I'll help you lots more than that, Ev!"

"Then you *believe* me?"

"Can't rightly say I actually believe you. It's a pretty wild story, but I'll help you find out if it's true. What I can do here in New Mexico, anyway. Can't go out more than a day or so, on account of my business. I'm a plumber, you know. Gotta make a livin'." He took another sip of his beer. "But I'll tell you, I don't *disbelieve* you. You see, Ev, that's my philosophy, and that's what got me in the saucer world."

"I don't quite follow."

"Well drink up that beer and this one, too, and maybe you'll get a little nearer."

Mashkin got up to get them both fresh beers. Scarborough shrugged and took a deep drink of the Budweiser. He felt comfortable in this house for some reason. He had the feeling that he could trust Mashkin, which was good, because he *had* to.

Mashkin levered off the tops of the bottles, split up the pair and handed one to Scarborough. "Here, you start working on this and come on down to the basement. I got something to show you."

Scarborough stood, took the beer and followed. "What's down in the basement, Walter?"

"You'll see."

He led him through a corridor branching off into bedrooms and a bathroom, and then opened a wood door that dived into gloom. He flicked on a light and beckoned Scarborough to follow.

After the comfortable but plainly old, lived-in, and rather tacky upstairs, Scarborough was surprised to find himself walking down the steps into a bright, well-lit, fully finished basement. One section was an office with a Leading Edge computer, a printer, book shelves, and the usual office accoutrements. Another was an entertainment center, including a large TV screen, a VCR, and a neatly filed stack of tapes. On the other side was a regulation-size pool table with neatly racked cue sticks and balls on the wall. Between these two sections was a large modern Magnavox stereo system, complete with tape deck, an equalizer, and, glory of glories—a compact disc player.

Scarborough wandered over to the shelf immediately above the system.

"Compact disks!" he said.

"Yep!" said Mashkin, beaming proudly. "Got a nice set-up here, don't I? Surprised you, huh? That there computer's what I do my correspondence and UFO articles on. God, I love to write! Nothing better than slipping in a good CD or tape and then letting these fingers go on my keyboard!"

"I know what you mean, Walter. I do indeed!"

There were a couple hundred CDs on the shelf, but when he looked at the titles, his heart fell.

They were practically all country-western and rockabilly. Absolutely no jazz. Nil classical.

Mashkin walked over and pulled out a jewel-case, holding it up for Scarborough to see.

"You like the King, Ev?"

The compact disk was *The Sun Sessions*.

"Ah, er—well, I'm not a big fan, no."

"Too bad. What a great singer. Whew. I remember, 1954, I was twenty years old and driving a truck down to Knoxville, and that voice of his comes warbling out of my car radio. Shivers just went down my spine. That's it! I think. That there's the voice of my soul And don't it just sound purty as hell, too. Well now, Ev, to get back to what we was talkin' about—belief and disbelief. You hear those stories about Elvis Presley still being alive?"

Scarborough cringed a bit.

"Well, you know, I don't know what to believe. I never met him, but I seen him do his shows. I seen him at Vegas, too, though that was just out of loyalty. But you know, I never seen him dead. And I know he's alive, here . . ." He pointed at his record collection. "Here." His head. "And here." His heart. Looking a bit somber, Mashkin then went over and pulled a sliding cabinet-door open. Inside deep shelves were piles and piles of tabloid newspapers. Scarborough glimpsed the *Sun*, the *Star*, the *National Enquirer*, and of course, the *Intruder*. Mashkin patted a pile contemplatively. "So I read all these reports. And they ain't just in the tabloids; I read the books about them, hear people's stories. I mean, of course, about the King bein' *alive*. What do you think, Ev? You think Elvis Aron Presley's fooled everybody and is walking the earth today?"

"I never really thought much about it, to tell you the truth," said Scarborough, diplomatically.

"You know what I think, Ev? I think, maybe it's true. That's my philosophy. I'm open. I'm what you call in your books 'credulose.' "

"Credulous, you mean."

"Yeah. I'm open. You know, since I been a baby, people been telling me things. They tell me things that later on I find out ain't necessarily true. They say, 'Walter! The Lord Jesus is Thy Savior and you're gonna go to hell, lessn' you know him as Personal Savior.' Well, heck, I like Jesus and God and all, and that could well be the truth. But you know, what about all these *other* beliefs? Are they all wrong? And they say, 'Walter! You gotta be married before you put your manhood where it was made to be put!' Well, long about eighteen years of age, that manhood of mine started leading me to where it wanted to go, and I didn't get hit by no bolts of lightnin' 'cause I weren't married to the girls. I just got some good lovin', some life, some friends, some enemies, in the process. 'Walter!' they tell me in high school science, 'There ain't no such thing as flying saucers!' But I'm twenty-two years old, driving a truck through the Smokies, and I swear to God, I see three of 'em, just asmoothin' through the sky like sailing ships of the night, lit up in just incredible colors." Walter shook his head, sipped his beer, pinky sticking out. "Yep, Ev. I'm a credulous son-of-a-bitch. And I'm damned proud of it, too."

"Have you seen other flying saucers, Walter?"

"Nope. Only time. It was enough, though. It was like Saul on the road to Damascus. Zap. And now, I'm just as fascinated with the damned things as ever before. Shit, we could talk a couple days about the subject, much as we both know about it, and not get anywhere. So why don't we just call a truce on it. Let's just say, Yep. I'll help you, Everett Scarborough, 'cause you're asking for help, and 'cause I want to find out what happened to Harry. And maybe, if you're right, and these bastards are doin' what you say they're doing, then to help stop it. Me and the government ain't real good friends, I tell you that!"

"I'm not disputing the system, Walter. I'm just saying that a group of people have apparently been successful in defying and abusing that system."

"Well hell, Ev. People have been doing that for years. C'mere, I wanna show you somethin'." Mashkin beckoned him over to another sliding door, and opened it. He moved some books and pointed.

Recessed into a cinderblock wall was a wall-safe.

"That where I keep my money, Ev. I don't got no truck with government. I do a purely cash business. I keep my money outta banks. They can track you that way, those IRS bastards. They can even steal the money away from you, if they want! Now, mind you, I pay the state taxes. I love New Mexico. And I love the United States of America, too, and I pay excise and sales taxes and all the other taxes. But Ev, if our Founding Fathers intended for American citizens to be taxed on their *income*, they woulda put that in black and white in the Constitution. Way I figure it, that part of the government is sucking us citizens dry. And let me tell you, they come snoopin' around here on my private-and-paid-for property, I got somethin' ready for 'em." He strode to another cabinet, this one on the wall, and flung the doors open. Hanging inside was a large collection of handguns, a rifle, and one shotgun. "Got another Remington in the back of my truck, too, and a damned good lawyer in the city, too."

"I see," said Scarborough a little taken aback by the fervor of this previously mild man and by his extensive gun collection.

Mashkin suddenly grinned and winked. "I just wanted to let you know, Ev, that I ain't got no compunction 'bout working against the rotten parts that have festered in this great nation's government."

"I'm afraid that you won't get a chance to shoot any revenue agents, Walter."

"Well then, CIA slime will just have to do!"

Scarborough shook his hand. "It's very nice to meet you, Walter Mashkin. I have the feeling that this will be a fruitful relationship for us both."

Mashkin's handshake was firm and cold from the beer bottle. "I'm glad, Ev. I'm real glad you called. Now, what say we play us a game of pool as kind of a sealing of the pact of friendship, eh? And then I 'spect you'd like to have yourself a little siesta, so I'll just show you to the guest room. That is, after I beat you."

"I don't know, Walter. I'm not too bad at pool."

"Ah-ha. I'd hustle you real proper-like, Ev, but I got a feelin' a man on the run from the government ain't got whole lots of money on 'im."

"You reckon right, Walter." Scarborough was honestly starting to feel good, at home, relaxed—the first sign of true lack of tension since he'd taken refuge in Marsha Manning's home. "But I guess I could spare five dollars to stake on the game. I could use another five . . . or ten, if you're so sure that you're going to beat me."

"Woo—ee. Is that a challenge I hear? Consider your five bucks gone, fella!" The lanky man strode over and carefully took the cover off the table, folded it carefully, and put it on a chair. He knocked on the green felt. "Genuine slate, Ev! Flat as Olive Oyl's chest! This thing's half a century old. I refurbished it myself."

"Very lovely."

"So." He gestured at the rack of capped wooden sticks. "Select your weapon! And prepare to do battle, as I rack the balls."

Mashkin happily set down his beer on a side table, put the rack on one of the table's two spots, and began to pull the balls from the pockets and set them into place. "Old-fashioned bottom, Ev. None of that chutes-and-ladders stuff, just some sturdy leather beneath the pockets! You wanna flip for break?"

"Better. I've never seen you break before."

Mashkin pulled a quarter out of his pocket. Scarborough won the toss. "Eight ball, I take it?"

"That's where you are, ain't it, Ev?"

"Yes." Scarborough lined up and shot. The break netted him the Seven ball, which meant that he had to sink the other solid balls before he tried for the Eight. "Behind the Eight ball."

"You know, Ev. I guess I was wrong all those years."

"Yes? About what?"

"Well now, I been following you for your whole career, UFO stuff bein' my avocation. And from the time I read your first book, first seen you on the TV, I would think to myself, Whew!—" He held out his cue stick held straight up—"Boy, that fella must have one of these things rammed up his rear, straight through his brain."

Scarborough thought about that a moment with a frown and a sigh, scratched his nose, and smiled sadly. "Well, Walter, when I traveled, I enjoyed a good game of billiards on the road and my bags were too small—so I guess I had to put it somewhere!"

As Walter Mashkin chuckled heartily, Scarborough aimed the stick against the cue ball and sank the Two ball.

He won his ten dollars handily.

CHAPTER TWENTY-THREE

The Gamma Complex, the core of the operation called White Book, was on the exterior a very plain and old-looking standard military building dating back to the postwar building boom and beginning to look its age. It stood about ten miles south of the center of Kirtland Air Force Base, harking back to a day of military expansion. Much of it had ceased being used much after the Vietnam War. . . but then, Brian Richards's predecessor had seen its potential, and established it as part of the White Book network. There were four buildings, lots of tarmac, and a helicopter pad, all surrounded by a wire fence with only one entrance and exit, where two armed guards checked in authorized personnel.

Despite the drab facade, Gamma Complex's interior had the streamlined look of a rocket-launching base, complemented by the latest in technological equipment gleaming behind stainless steel doors. However, as modern as the computers, the lab equipment, and the medical paraphernalia employed were, the conference room that Dr. Julia Cunningham sat in now was just as drab and spartan as the quarters they'd assigned her in the accommodations building to the north of the quadrant.

It was four o'clock in the afternoon, and they were supposed to be here, dammit!

Cunningham was a person who valued punctuality, valued her time schedule, and certainly had pressing duties ahead of her in the lab. They'd already postponed the morning meeting, and rescheduled it for four P.M. so now where the hell were they?

She poured herself another glass of the iced mineral water and sipped. She felt much better now, much improved after her ordeal of the morning. It wasn't just the drugs she'd added to her neuro-chemical network. A good day's work, with a nutritious breakfast and healthy lunch, had evened her disposition out, and she was starting to feel normal.

Up until now, that was.

As she again examined the digital face of her watch which read 4:05:24, she realized she had the beginnings of a headache.

Damn Richards!

Logically, she knew that there must be some kind of glitch, but somehow she could not help but feel that the man was doing this to *punish* her, *taunt* her. Although their sexual relationship had been over for several years, after they agreed that it was too dangerous and might foul up the operation, Cunningham could not help but continue her emotional relationship with him. She still loved the bastard, in a deeply disturbed and twisted way, and she suspected why. A shrink would tell her that it was because of her sick relationship with her father.

Somehow, she was drawn to men who mistreated her. Nonetheless, try as she might to destroy the feeling with her medications, her own designer-chemical therapy, she could not shake it. She thought of Richards all the time. She still wanted him. Down deep, she prayed that, after White Book was completed and in a maintenance phase, when they would be out of danger, Richards would divorce his wife, leave his family in Great Falls, Virginia, and come and live with her.

She knew it was sick, this feeling, this need, this desire, and she had tried every means possible to cure herself; but the feeling was still there.

Nonetheless, she was in *control*.

She kept rigid watch over herself, lest this aberrant, pathetic side of her might show to anyone. And she was successful. However, at times like these, when things got out of hand . . . When she lost Woodrow Justine . . . After Scarborough escaped into the wilds of America, a walking time bomb teetering on the truth about himself and about the operation . . . These were the times that things got rough. These were the times when, perhaps, she tended to overmedicate herself and get the headaches.

She did not take an aspirin or a Tylenol. That would be an inappropriate mix with the other chemicals in her system. Biochemistry was never as exact a science as Cunningham would have liked, and there were so many random factors. However, she was working on that . . . This was one of her own private reasons for helping to develop Project White Book. There was so much she was learning about the way the brain operated—and without petty moral or social constraints.

Ultimately, Dr. Julia Cunningham knew, given the proper conditions, she would be able to control the heart, the mind, the soul, of any man or woman.

Perhaps even herself.

It wasn't until 4:12 that Brian Richards walked into the conference room, bringing with him a youngish-looking man with blonde hair and a deep frown and a resigned look to his eyes. By that time, Cunningham's head was pounding. But she wouldn't give Richards the satisfaction of knowing what he'd done to her.

"Oh," she said casually. "Glad you could drop by."

Richards cleared his throat. "Yes. Sorry we're late. Myers's plane was late, and then we had to get ourselves a copter from the base."

"You couldn't drive over."

Richards ignored her. "Ed, this is Dr. Julia Cunningham. She's our chief scientist for Project White Book, some of which we'll be showing you today."

"Doctor," said Myers, nodding.

"Hello, Mr. Myers." Cunningham replied tersely. "I understand that you're being a great deal of help in the present situation."

"If you mean that I'm the Judas goat . . . well, I guess you're right."

"I understand that you are not happy about the circumstances. Nor would I be, in your situation. However, I hope you have been convinced of the terrible importance of the operation."

"Yeah. I figured I better be convinced," said the CIA operative, giving Richards a dark and surly look. "Or else."

"Ed has been given an outline of the basic information. However, since there's still some time before an—er—appointment can be made with Everett Scarborough—"

"You mean 'trap,' don't you, Richards?" Myers slumped into a chair and moodily poured himself a glass of the water. He drank it down in several quick gulps.

"Very well, then. Trap. Whatever you wish to call it, it shall be the means whereby we can get Everett Scarborough in our hands. I explained to Ed, that by doing it this way, we may not have to kill Dr. Scarborough—it could be possible to merely persuade him to change his mind."

"You mean brainwash the poor guy." Myers poured himself another glass of water.

"I thought that by showing Ed our facilities here, with your able explanations, Doctor, that he might come to see that our work is not only in the best interests of the national security . . . It is, above all, philanthropic, humanitarian—in short, for the good of mankind."

"Well, I'm not sure if I can give these kind of public relations speeches, Mr. Myers," said Dr. Cunningham. "But as Mr. Richards *is* my boss, if he clears you, I'm quite willing to show you the labs and talk about our work here, and other places around the country—and, through the Publishers, the world.''

"Yeah. I want to hear all I can about these Publishers."

"That depends on what Richards clears me to say."

"You have to understand that Dr. Cunningham's knowledge is kept limited as well. But I don't see any harm in you telling Ed everything you know. I think, by the time you're finished, we may well have convinced him that the Publishers are perhaps the most important group of people in the world today...A group that has helped mankind survive this far ...And may well be responsible for preserving the race through the troubling times ahead."

"Christ, by killing people? By betraying friends? By threatening to destroy people's families?"

"Ed. You must know that true leaders always follow John Stuart Mill's dictum: The greater good for the greatest number. I regret what we have had to do in this case, I truly do, but you gave us no choice. Everett Scarborough is now a great threat to our program. He *must* be captured."

"Well, I'm doing what you tell me. Isn't that enough?" He drank more water. The spring sun and heat in New Mexico must be having an adverse effect, thought Cunningham, automatically calculating the necessary chemicals she might administer. She stopped herself. There were more pressing things to think about than a CIA operative's hydrovascular system. Still, to Dr. Julia Cunningham, everything was chemistry.

Everything.

"I'm sure you are doing your professional utmost," Richards responded. "However, Everett Scarborough is not a stupid man. Somewhere along the line he may well perceive a tremor, a lick of the lips, a gulp of regret. Mere words on paper do not transmit the full enormity of the situation. I am hoping that your tour of these facilities, as well as the explanations from Dr. Cunningham, will give you the conviction your act needs."

"It sounds as though it's going to make me *sick*."

"We shall discover that in due time, Ed. Now then, Doctor. Would you like to begin our little show-and-tell for our esteemed colleague?"

"This way, please, gentlemen. We'll start just down the hall, toward the true core of our efforts here."

Dr. Cunningham stalked toward the door.
Her head was feeling immensely better already.
She enjoyed lecturing people.
She felt in control.

"It's a computer."

"Yes, how observant of you, Mr. Myers. However, it's a very special computer, a very *unique* computer, developed specifically for Project White Book—most particularly for my analytical work on the operation. It performs a variety of functions gathered around the essence of our work here: the practical study of methods to shape the very fabric of human existence."

They were Cray computers, many of them, with screens and readouts and printout boards connected with multiple cables and peripherals.

"Mind-washing again," muttered Myers.

Cunningham shook her head adamantly. "No. Brainwashing doesn't really work except in the most extreme of circumstances. The human mind is a complex organism, and must be treated as such. Brainwashing is ineffective, dangerous, and messy. It leaves holes as wide as barn doors in the behavior of individuals. And frankly, it harms them. Ultimately, we do not mean to *harm* people. Or even control them, actually—just *mold* them. Social herding. But I get ahead of myself, Mr. Myers. Richards here wants me to tell you about Project White Book, and that I shall do. I just want to impress upon you that this is not the primitive torture ritual you might think it is."

"What, you work with subliminals? I thought those didn't really work either."

"Subliminal programming *is* an element of the mix. As is hypnosis and what we call "Reality Programming." But above and beyond all this, there is chemical bioengineering."

"You're getting ahead of yourself here, Doctor," said Richards. "Myers seems very upset about our brief borrowings of individuals to work with them. Maybe you'd better take it from there."

Faintly annoyed that she had to cover even the most obvious basics, Cunningham sighed and said, "Very well. I trust that by now, Mr. Myers, you are familiar with the basic history of Unidentified Flying Objects in the U.S."

"Yeah. *Too* familiar."

"The actual phenomenon itself—the physical evidences of bright lights in the sky, etcetera—doesn't really concern me. What interests me is the phenomenon of what goes on in the human mind during this kind of exposure to the unknown . . . and what it is that *exists* in the mind that can actually project this kind of experience when in fact it is only a subjective reality."

"You mean a delusion."

"Exactly. Why is it that when a few people claim to see flying saucers, *many* people suddenly see them? A form of mass psychosis. Was Carl Jung correct? Does there exist a collective unconscious? But in fact is it *more* than the mere *tendencies* toward archetypal influences? Is there really some kind of psycho-chemical connection between people? But again, I get ahead of myself. Let's get back to UFOs, shall we?"

"Yes, let's do," said Richards, clearly a little peeved at all this aerie-fairie pontificating—just like Cunningham knew he would be, which was the reason she did it.

"In the early sixties, the UFO psycho-cultural experience changed. A couple named Betty and Barney Hill were driving home from a vacation toward their New England home when they saw a flying saucer. They stopped their car. And then they apparently lost consciousness. When they got home, the couple realized that they had lost several hours of time. What had happened?

"Under the ministrations of a psychologist and the use of hypnosis, the Hills started to 'remember' what had happened. They had been abducted, they believed, by little men from a flying saucer, taken upon the craft, and there were subjected to a battery of tests. They were then told that they would forget the experience, which was a deeply frightening one, they claimed, very *protean*—my word, Mr. Myers, and may I add the aforementioned term *archetypal*, as well."

"Well, did it really happen?"

"No, of course not, but that makes no difference. It was in the air, Mr. Myers. If it hadn't happened to the Hills, it would have happened to someone else. All the mass hysteria about flying saucers was only the crest of a deep wave, particularly in

Betty Hill's case. Do you think that she was shut off from the massive amount of science fiction, in film and television, as well as in magazines and books? There was nothing really original or startling in the idea of an alien race examining human beings. Maybe Betty Hill read the books of George Adamski, or just a particularly frightening science fiction story; and then, to top it off, saw *Invaders from Mars* or even more probably *Earth Versus the Flying Saucers*. The creatures from that particular movie clearly influenced the general view of what aliens might look like! It was all in the air, as I said. It was under people's skins, and I have no doubt that there were millions of similar dreams in the bedrooms all across America—dreams later forgotten, as dreams usually are.

"A couple of years later, Project White Book discovered this incident, and my predecessors saw its incredible potential to not merely serve their own purposes, but to experiment with psycho-cultural influences. I won't go into the details, and of course you know that the White is basically a cover-up operation concerning UFOs, underpinned by a program of disinformation—but essentially, we arranged for a book to be published concerning the story of the Hills, and had it excerpted in a wide-circulation magazine the name of which I can't recall . . ."

"*Look*," said Richards.

"Yes, that's right. *Look* magazine. *The Interrupted Journey*, by John Fuller."

"Correct. The effect of that book took a while to dribble down to the general unconscious level of the American public, but pretty soon, *other* people started having these kind of experiences. They felt they had been abducted by aliens in flying saucers. Our resources checked this extensively. There were no indications that this was actually happening. And yet, just by suggesting that it had happened and would happen again, people all over the country began to experience it as a kind of hypnogogic reality. We were witnessing a very real shift in the nature of the cultural *zeitgeist*—a shift that White Book had helped along! But not only that, much more significantly, people began to have entirely *different* attitudes toward the types of creatures that might be operating these vessels from different worlds! No longer were they at all friendly or even detached—suddenly, they were very threatening.

"White Book knew they were onto something. They'd already of course created false UFO-sightings from the very beginning, and now they had another way of spreading disinformation."

"This is why they began to kidnap people and make them think they'd been picked up by saucers?"

"Yes, the principle reason. It was never an extensive operation. As strides in science were made, so strides were made in the sophistication of the operation."

"And naturally, we made a giant step when we employed the good doctor here," said Richards.

Cunningham coldly ignored him. "This is not a fascist program, Mr. Myers. We carefully select the proper individuals, most likely to ably pass on their stories, disseminate the information amongst the culture. Typhoid Mary types, if you know what I mean. Persuasive and yet maybe a little off-the-wall. We don't want this to be a massive phenomenon. Just suggestive . . . I realize it's an odd kind of subtlety, but so far it is working very well. Which brings us to the next station of our tour. If you will follow me, please . . ."

She spun about, and marched away. The two men followed her.

"Here, Mr. Myers, is some of the equipment that we use."

The room was filled with odds and ends of paraphernalia. She opened a closet to show him the rack of alien suits, along with the extensive makeup kits utilized.

Myers walked over to a large machine which looked like a high-tech film projector. "Hypnosis device, I take it."

"Yes."

"And so you stage these abductions, examinations . . . what have you . . . here."

"And in other Safe Stations through the country. The environments vary according to the nature and intensity of the abduction experience desired. Let me show you one of our fancier stages. This way, please."

They walked down to a room with a lock that needed Cunningham's ID. She slotted her card, and the door clicked open. A *shush* sounded; the room had been slightly pressurized. Myers noted this.

"Yes, that creates the desired effects on the patient—that sense of being in an aircraft or a spacecraft. Purely associational, of course. The only experience the average person has of that is passenger jets. We create that kind of environment, although significantly enough altered to appear alien. Come on in, gentlemen."

They waked into a small room that had on the walls a multitude of gemlike encrustations, which seemed to be a curious coppery metallic alloy. The angles of the room were of a different geometric design than human rooms, and the ceiling, containing clusters of lenslike devices peering down onto the floor like eggsacks, was tilted at a disconcerting angle. The entire effort was one of *otherness*. It looked like something out of a science fiction movie designed by Bosch.

"Christ, this could be from the planet that the creature from *Alien* came from!"

Richards coughed and smiled knowingly at Cunningham. "Let's just say, Ed, that White Book has had some significant input in Hollywood, financially and otherwise. A terrific source of cultural manipulation."

"Although not, I take it, with Steven Spielberg?"

Richards immediately frowned. "No. Although we cannot prove it, we feel that Spielberg's influences are quite different."

Myers walked to a wall and touched it. "Yes, yes, this is quite good that you showed me this. And I take it that there's a lab someplace. Beakers? Tubing? A bubbling cauldron, perhaps."

"I can show you the lab if you like, although it's just a lab, albeit quite advanced," said Cunningham.

"Advanced. Right. Gotcha. So you haul these poor people out of their homes in the middle of the night, after you've selected them for their big mouths. How the hell do you do *that*?"

"Again, with the use of hypnotics, subliminals—but most important, we introduce a highly effective gas into their homes that places them into a highly pliable form of unconsciousness," said Cunningham. "That was developed in the seventies by a separate branch of the CIA working on chemical weapons and nerve gases. As soon as White Book saw its potential, they began to use it."

"And you haven't gotten caught by local authorities? Surely you've slipped up from time to time."

"Oh, indeed," said Richards. "Sometimes it is necessary to actually deal with certain people biochemically to relieve them of their memories, but without the introduction of the alien-abduction motif."

"An awfully dangerous weapon! What do you use it for, to change people's minds about whom they're going to *vote* for?"

"A far too extensive proposition. No, we're interested in politics only to the extent that it affects our operation. The Publishers, of course, are most definitely suprapolitical. Let's just call us Editors their executive branch."

"The Publishers. The fucking Publishers!" Myers shook his head and leaned against the special-effects wall. "You know, I'd heard rumors about them ever since I made high-grade. But I always thought that it was all just paranoid conspiracy bullshit, you know? Robert Ludlum territory!"

"I am unhappy to say, Ed, that Mr. Ludlum is not in our employ as a very excellent disinformant to the popular culture. His delicious and wild enclaves of powerful people are nothing compared to the truth—however, we haven't discouraged his sales. Fictional camouflage."

"Wheels within wheels. But I *still* don't understand. Just who *are* these Publishers?"

"Their identities, of course, we cannot tell you. However, we can tell you a little bit more about why they exist and their highly important role in the history of the United States—and all mankind, for that matter. But perhaps we could talk about all this over a cup of coffee. This late in the day, I myself could use a bit of a pick-me-up. And Dr. Cunningham here—well, she's always in the mood for a pick-me-up, aren't you Doctor?"

Cunningham glowered and stalked away.

"Come on, Ed. We'll order some fresh-brewed French Roast in the conference room. And I'll tell you a little story, hmmm?"

Richards ushered the frowning CIA operative out the door.

The walls glittered behind them like a fabulous treasure cave from a story from the Arabian Nights.

And then the lights went off.

CHAPTER TWENTY-FOUR

With a rush of air, the seam in the bottom of the landed flying saucer cracked open, revealing the outlines of a door in the shiny alloy of the ship's side. The door opened outward, slowly changing into a ramp. A mist backlit by swirls of aquamarine and crimson washed down onto the desert ground, folding over the feet of the man, the woman, and the child looking up at this vision, transfixed with wonder.

The ramp clumped against the ground.

Within the fog, something moved.

Something began to walk down the ramp.

"Daddy, Daddy!" cried the little girl, pointing, eyes alight with curiosity and yet fear as well. "Look!"

A voice bellowed from behind them, "Cut! Okay, that's a wrap for the morning! Break for lunch!"

As the director waded through the colored dry-ice fog to talk to the actors for a moment, and the gaffers, best boys, boom operators, and cameramen secured their equipment in order to take some time off to get something to eat, Jake Camden tapped the shoulder of the man who sat in his director's chair, watching the activity around him impassively. "Okay if we talk *now*, Max?"

Maximillian Schroeder turned and regarded Camden for a long moment before showing recognition. "Ah yes, Jake. I'm sorry to keep you waiting so long, but this is a crucial scene and I wanted to make sure that all the details are correct."

Jake Camden nodded, looking around him at the movie set just outside of Sante Fe. The "flying saucer" was, of course, only a partial construction. Walk a few paces to the left, and you'd see all the special-effects people operating the lights and the mist and what-have-you. The ground and surrounding rocks were real enough, though: this was a location shoot, and the surrounding scene fairly bristled with high-tech equipment of the moviemaking kind. Yes sir, thought Camden. This was no low-budget grade-B film. This was top-drawer stuff. Schroeder's first film had been a box-office bonanza, and his film company had the bucks to make its sequal look really good.

"Well, I must say this is a far cry from the movie that I consulted on. They shot just about the whole thing in a warehouse in Pasadena. Exteriors in Griffith Park."

"Hooray for Hollywood, eh, Jake?" said Schroeder, smiling for the first time all day. "Even with script, director, and overall final approval, I still have to be here to make sure they don't make any costly mistakes or changes. I must admit, though, I didn't expect you to trundle in unannounced bright and early this morning in the middle of New Mexico!"

"I really have to talk to you. Privately."

"Obviously." Schroeder touched the blue and green rayon of Camden's Hawaiian shirt, today featuring sailboats supporting bikini-clad woman. Camden had not had a chance to wash clothes for quite awhile, and the shirt looked quite rumpled. "You're looking a little tatty, Jake. Everything okay?"

Camden eyed the surrounding film crew, most of whom where heading off to a number of tables beneath tarpaulins to pick up sandwiches, Cokes, and coffee. "No, Max," he said *sotto voce* from the side of his mouth. "That's why I've been dying out here, trying to keep a low profile waiting for you."

"Ah. Well, then perhaps we should take lunch in my personal trailer, hmmm?" Schroeder smiled and pointed to an area past a rock grouping. For his own part, Maximillian Schroeder dressed in what might be considered desert sartorial splendor. He wore khaki shorts, a khaki shirt with epaulets, and a bright white hat atop his well-groomed head. He smelled of expensive Calvin Klein cologne. He looked tan and healthy, not at all like a person who claimed to have spent a good deal of time on the examination table of bug-eyed aliens.

Camden smiled and relaxed. "I was hoping you'd have one of those. The film I worked on, my personal trailer was a Johnny-on-the-Spot!"

Schroeder shrugged and gestured for Camden to follow.

Maximillian Schroeder's trailer was actually only a third of a long trailer, but it was rigged with all the amenities. Wet bar, refrigerator, comfortable chairs, private bathroom, and cellular phone. It had air-conditioning too, which wasn't really necessary, since it wasn't all that hot outside and dry heat at that—but Camden just plain enjoyed the whir and cool of ACs, so it put him further at ease. After he'd used the can and accepted the long Tom Collins that Schroeder fixed for him, he settled on a pillowed couch and let out a long "Ahhh!" of relief.

"So, what trouble are you in this time, Jake?" asked Schroeder, not one to mince words.

"Florida State Police, I think. Definitely the FBI. I'm no longer working for the *National Intruder.* I'm running out of money. And did I mention the CIA? Yeah, I think they want me, too—either them, or one of their illegal branches."

The ice tinkled as he took a cool sip. The citrus was nice, but the alcohol was nicer.

"My goodness, Jake. You've truly outdone yourself. And all of this because of your involvement with Everett Scarborough?"

"Well, part of the problem—the police and FBI and money part—has a lot to do with drugs, so you could say a large part of it I put upon myself."

"But this illegal CIA branch . . . That's something you intimated before. Perhaps you'd care to elucidate."

"Look, Max. First, I gotta know if you're behind me. I gotta know that I can count on you to help me out. You say you're interested in film rights, but I haven't seen a thin dime yet. You helped me out with the agent—he's got lots of nibbles on the story, and I'm pretty sure that all this is going to turn into not only a gold mine but the success story of the decade. Right now, though, I'm broke and I'm in trouble . . . and I need help. So, can I count on you?"

Schroeder got up and walked to a desk on the other side of the room. Sitting there was a Gucci leather briefcase. He opened it and pulled out a packet of legal-sized paper, stapled in the upper left-hand corner. "I heard you'd called, Jake, but I didn't think the situation was that pressing. I assumed that you'd call here. I didn't think it would be in person, although

I'm not unhappy to see you. I wish you the best with the articles and book—alas, I can't help you with those more than I already have. However, if you'd just sign these, I believe that the movie deal will be taken care of."

He handed Camden the papers. Jake glanced them over, and grinned. "This is an option contract. Twenty thousand dollars renewable in a year... a hundred thousand if you buy... points... Not bad, Max, but I'm sure you can do better."

"I just so happen to have two thousand dollars cash on me right now, which will be the first part of the advance. Of course, I'd be willing to negotiate with your... *our* agent. But I thought you needed money quickly, Jake."

"Where do I sign?"

"Usual place. At the end." Schroeder drew out a gold ballpoint pen from the briefcase and handed it to Jake, who immediately used it to sign his name three times. When he looked up from his signature, Schroeder was opening his billfold and withdrawing a number of hundred-dollar bills, which he counted out on the coffee table. Jake counted them again and then happily pocketed them.

"Thanks, Max! You won't be sorry."

"I hope I won't."

"Well, this helps me out a lot, cashwise. But I'm not going to be able to write my article or my book unless you help me out the rest of the way. Looks like you've just made an investment in Jake Camden, buddy, and thanks." Jake grabbed Schroeder's hand and pumped it. "Now you're going to have to protect that investment."

Schroeder parted their grip with a look of mild distaste. "Jake, you really don't have to play salesman with me. I'm well aware of my present responsibilities." He went and poured himself a glass of Perrier with lime. "The question is, are *you*?"

"Huh? I don't follow."

"If you had taken the time to read that contract fully, you would realize that we have formed a partnership of sorts in regards to this operation. You seem to also forget, Jake, of my own interest in the UFO conspiracies and such. I need to be kept up to date on these things—I hunger for information.

Like, for instance, where is Everett Scarborough now and what is he doing?''

"Shit, I can't tell you everything! I mean, I know I can trust you, Max, but all the same—it might slip to someone else, and then where will Scarborough and I be?''

"Then you *know* where Scarborough is?''

Camden took a long drink from the tall glass, pondering. Why was Schroeder so damned nosy about where Scarborough was? This was rather disturbing.

Of course, he knew that Scarborough was visiting Walter Mashkin not far away, in Albuquerque. He'd remained in contact with Marsha Manning, getting information from her as well as giving it. Also, Marsha was planning to come down to Albuquerque. Something about helping Scarborough out, as well as investigating that hoary old business of the crashed saucer in Roswell. Sheesh, and hadn't he milked *that* one dry over the years! Ha! Schroeder didn't know anything about Manning, and that's the way it was going to stay. But Scarborough was a different matter. Scarborough was the *subject* of this business, this book, this movie. What harm would it do to give some information to Maximillian Schroeder, a man who was almost as much an anathema to the government for finger-pointing on the subject of conspiracies as anybody. And he'd also *certainly* be interested in hearing about the strange things that had been happening in regards to the strange guardians who'd been tailing them and protecting them . . .

"It is, after all, the next part of the plot,'' Schroeder prodded.

"True. And maybe you can help us.''

"I'll do anything and everything I can, Jake. You know that. Now that we're partners, I guess I'm obliged to.''

"Well, I can't give you specifics. Not that I don't trust you, Max.''

"That's all right. I understand. You have an important person to protect. Generalizations will do quite well . . .''

"Okay. He's in New Mexico. We've got reason to believe that the base of operations for this CIA branch is quite near Kirtland Air Force Base.''

"You don't say.'' Maximillian Schroeder wore a veiled look. "But operations you say, Jake. What are you talking about? To cover up the contacts that the Others are making with human beings?''

''Look, Max, I know this may come as a shock to you—but Scarborough and I . . . we've got reason to believe that a good part of this alien-abduction thing . . . Well, it might be a government conspiracy. For some reason, the CIA is kidnapping people and making them believe they've been on alien spacecraft. Don't ask me why but—''

A sudden realization occurred to Camden that stopped him in midspeech. Jesus! He hadn't thought this thing all the way through—Oh, no! If these bozos were manufacturing fake abductions, then Max Schroeder could have been one of their victims! Which meant that maybe *he* was being watched as well.

''Go on, Jake.''

''Uhm . . . well . . . maybe I should just go, Max. I mean, I don't want to get you *too* involved! It could be very dangerous.''

''Jake, you're not telling me something that I haven't heard before.'' Schroeder stared at him, dead-earnest. ''The Others—they've told me about this occurrence. They are very upset. I have not mentioned this in any of my books. It is too confusing at present.'' A troubled, almost haunted, look seemed to invade Schroeder's eyes. He turned away for a moment, stiffening, his face undergoing a remarkable transformation. For a fleeting moment, he was no longer the suave and collected Yuppie minister of the New Age religion of the Others, but a person in fear and doubt and pain. A paroxym of angst washed over that face—and then, just as suddenly, he gained control again. ''Forgive me, Jake. I—when the memories of my contact with the Others hit me full, as they just did, I am not myself.''

Jake Camden's mind was going a thousand miles per hour. ''Hey, maybe you'd better have something a little stronger than fancy French soda water, huh?'' Geez, there were tears in the guy's eyes!

''I don't usually indulge, but in this case . . .''

''Here, you just sit here and I'll get you something. What, some sherry, something watered down?''

''I believe there's a bottle of Glenlivet Scotch in the cabinet.''

Jake found it. ''Water? Ice?''

''Neat.''

"Wow. I'll drink to that!" Camden poured two healthy tots of the amber liquid into tumblers. This really had the guy shook up, Camden thought. What's more, it really seemed *genuine*. Jake had always considered Schroeder a bit of a loony, but now that reality was getting a bit out of kilter, he thought that maybe old Schroeder here was on the up-and-up about these suspicions of CIA-disinformation abductions.

"Here you go." Jake handed the man his drink, and Schroeder took a gulp.

"Thanks."

"Help?"

"A little."

"Okay. This is *real* interesting, Max. About the Others knowing about the CIA abductions. I mean, the thing is . . . And pardon me for doubting, but did it occur to you that maybe *you* were the victim of some kind of brain control?"

Schroeder shivered. He took another drink. Shivered again. "Yes, Jake. Don't you think this whole thing hasn't haunted me from the beginning? Incredible paranoid permutations of the situation have run through my mind from the beginning. Many, I've included in my books. But I've not included the CIA participation the Others told me about—for fear of my life."

"What have you done about it?"

"Well, I've talked with the Others, and they agreed I should attempt some kind of discreet inquiry. As you probably know, Jake, I come from not only a wealthy family, but a well-connected one. I know many people in high places. I inquired. I found absolutely nothing to indicate this kind of CIA activity. However, I *have* cultivated relationships with certain members of that organization whose honor and integrity are above reproach. I am *positive* that they could not possibly have anything to do with such a heinous flouting of individual rights."

"What, you don't think the Others flout your rights by dragging you on their spaceship?"

"They are not bound by the laws of this nation. They do not truly understand human emotions. This is the hellish part of the contact . . . Most are not even concerned that the CIA have mimicked their procedures—in all cases, distorting the experience so that the spirituality and the emotional transformation involved are entirely turned into a totally *negative* experience. There is much that is heavenly about the relationship between the Others and their human contactees, Jake. But the experience that this group is creating is absolutely *hellish*. It negates everything that I have worked for in my books and my movies! Naturally, I want to end this atrocity. But how?"

"You say that you have contacts in the CIA?"

"That's right."

"You know anything about the activities out there on the edge of Kirtland . . . that lab or whatever it is?"

"No. I wasn't even aware that it existed."

"Do you think that you can call your contacts and find out? Maybe arrange for us to visit or something?"

"Well . . . if it's a classified area, that's pretty hard to do."

Camden stared him in the eye. "Come on, man! We're working for the Others, here! These CIA people are besmirching the good works, the *cosmic* efforts toward the evolutionary development of mankind! We're talking the *future* of the human race. This might prevent nuclear war or something good like that."

Schroeder nodded. "Maybe you're right, Jake. I'll make the calls tomorrow."

"Tomorrow may too late. You've gotta do it now. Look, it's twelve-thirty here, which makes it late afternoon in Washington. Make the calls."

Maximillian Schroeder leaned his face into his cupped hands as though they were an oxygen mask, took a deep breath, and then slapped them down onto his knees. "All right, Jake. I'll do just that!"

"Good for you." Jake made a congratulatory salute with his drink and then downed it. He was feeling better already, much better.

"Yes—but, er, Jake . . . maybe you'd better leave me alone to make the arrangements." Schroeder walked to the desk, scribbled something on a piece of paper, and handed the note to Jake. "This is my personal phone number. Even if I'm on the set, I'll get word and I'll come over here and take the call. Maybe you're right, Jake. Maybe we'll be able to see just what is going on in the Kirtland facility. Yes. The more I think about it, the better I feel."

Camden pocketed the paper. "Great. I'll call you first thing tomorrow morning."

"Where will you be staying?"

"Don't know yet. Probably a hotel in town."

"Fine. You can give me that information tomorrow."

They exchanged meaningfully sincere looks, shook hands, and Camden was off, the money feeling solid and good and his pocket, the future looking much, *much* rosier.

As soon as Jake Camden was gone, Maximillian Schroeder was on the phone.

"Gregg? Yes, it's Schroeder. Gregg, I presume you saw the rather seedy-looking fellow hanging around on the set today and going into my trailer. Yes, that's right. He came in a Ford Escort. Gregg, he just left, and I want you to follow him. Report back to me where he goes, what he does, and whom he talks to. Understood? Fine, then you'd better be off because he's probably almost to his car now."

Schroeder cradled the phone, smiled, and went to get himself another scotch.

CHAPTER TWENTY-FIVE

They were back in the conference room.

They had ordered their coffee and now they were drinking it. Edward Myers drank his with cream. Richards drank his black.

Dr. Julia Cunningham stuck with her mineral water.

She sat straight in her chair, listening and participating in the conversation with an alert detachment. Actually, she would have preferred to have been back in her lab, working. That was her universe, not this administrative stuff. That was Richards's department. From time to time, she might be called upon to make pertinent contributions, which was why she listened. Otherwise, her mental facilities ranged back to her work, running over a particularly thorny recent neurological puzzle, like fingers clicking on an abacus.

"The Publishers," said Myers. "Go ahead, Richards. I'm listening. Now that I'm working for them, albeit against my will, I guess I should have a slight understanding of that organization."

"No, you don't understand, Ed." Richards sipped his coffee thoughtfully, looking out at the beautiful golds and browns on the ragged New Mexico plain beyond the electrified security fence. "We're telling you this because I sincerely believe that once you understand that this is a philanthropic organization I belong to, a *humanitarian* effort, you will be a happier camper."

"Yeah. Sure."

"The Publishers. Now, where shall I start?" Richards tented his fingers, pursed his lips contemplatively and then began. "Well, since we shall have to be very general for security purposes, very vague . . . we shall say that the Publishers began well back in the nineteenth century, at the crest of the wave of industrialization, when progress took the power of the British Empire around the world—and huge fortunes began to be accumulated by the enterprising capitalists of this great nation."

"They began in America then?"

"Yes, that's right. But they had influential contacts around the world. I have reason to believe that there are traces of Masonry involved, but this is by no means a religious sort of thing. Essentially, many of these rich and powerful men began to perceive the need for a suprapolitical organization to watch after not merely America's welfare—but more specifically, its national identity, its cultural integrity."

"You're talking about people like John D. Rockefeller . . . Andrew Carnegie . . . that lot."

"Please, Ed. I can't mention names."

Myers grunted and got a knowing look to his face. Cunningham knew that he'd already started to draw conclusions, not that it made any difference.

"Cultural integrity . . . Ed, I'm repeating that word because it's important. And I'm not necessarily talking about culture, à la concerts and art galleries and literature and such, although they are part of the mosaic. I'm talking about language, customs, national identity. . ."

"Race?"

"Precisely."

"We're talking about a bunch of Nazis here?"

"Oh, no, no. You're taking this the wrong direction, Ed. Quite the opposite. The Publishers are not purists. They are well aware of the contributions of all races to the wonderful melting pot that is America. Presently, practically all races that comprise American life are represented by members of the Publishers. You must remember that capitalism is the cornerstone here. Just as in America there are rich blacks and Latins, Orientals and Irish Catholics, so too, are they members of the Publishers."

"It's some kind of supragovernment, then."

"Not government, Ed. *Watchdogs. Guardians.* And arbiters of national direction."

"With an enforcement division."

"Wedded with the nation's executive branch, and therefore quite legal within the tenets of a visionary constitution."

"In other words, you're stretching the definitions of government limitations into a different dimension entirely."

"Over two hundred years have passed since the forefathers drafted that brilliant document. Some of their descendants are members of the Publishers, as a matter of fact . . . However, time has changed the world. It used to take a year to travel around the globe. Now the shuttle circumnavigates it in a couple of hours. I need not tell you that it is a weary and troubled world that we occupy. Forces of progress hurtle on at tremendous rates, far too quickly for our feeble selves to truly keep up with. Quite simply, the Publishers exist to preserve America. They continue the dream. Just as our defense system guards the borders of our land, so the Publishers guard the borders of our national consciousness, unconsciousness—and, if you'd care to borrow a Jungian term, the American *collective* unconscious."

"You mean, like advertising and public relations and religion, only on an almost cosmic level."

"There's nothing 'almost' about it." Richards smiled, looking terrifically pleased with himself.

"Whew. This truly takes the notion of conspiracies to that level as well."

"The Publishers and Editors do not consider themselves a conspiracy. We do not act for private gain. We ultimately act for the welfare and continued future of our great nation."

"Wait a minute . . . You must make a bundle. And you're *bribing me*!"

Richards raised an eyebrow. "Membership requires great personal and emotional risk. There should be some paltry reimbursement."

"Okay, okay. I've got it. You're the flack gods of America. But *what* does this outrageous business of *kidnapping* people and then using *mind-control* methods to make them think they've been abused by aliens have to do with preserving national integrity?!"

"I would have hoped you would understand. You read what I gave you, true? You realize that the U.S. government and military have been covering up the truth about UFOs almost from the very beginning?"

"Sure, sure. You know, I'd bought the company line on that, too. I thought that back in the forties and fifties we were just worried about the Russians pulling some double whammy on us. We were real jumpy about national security. And then, as usual, we just didn't want to admit that, so we brushed it under the rug and put a stonewall around it. I didn't realize that we were so freaked about people thinking that we'd lied to them that we'd try to divert their attention with a bizarre program like this. What is it? Make the UFOnauts so freaked about abductions that they're so busy watching the skies and interviewing and hypnotising people, that they forget the U.S. government's part in it from way back when?"

Richards turned to Cunningham who regarded them both with statuelike placidity. "I'm afraid I didn't include the big picture in the folder concerning Project Black Book," he sighed. "Now that Mr. Myers has proven his worth and value—and loyalty to us, I might add—I feel that perhaps it's time to give him the bottom line."

"What, the truth?"

"All you've heard today is absolutely true," said Brian Richards, his face grave, and looking much older for it.

"We've just left out perhaps the most important part—it least of *this* section of—uhm—Editorial Operations."

"All right. I'm a big boy. Please let me know why I'm betraying my friend," said Myers, the bite of sarcasm still plain in his voice, but his face showing honest curiosity.

Richards nodded to Dr. Julia Cunningham. She smiled, relishing his sense of the dramatic, and turned to the CIA operative.

"It's really quite simple, Mr. Myers. You see—the United States of America really *does* have extraterrestrial visitors."

CHAPTER TWENTY-SIX

Everett Scarborough was helping Walter Mashkin carry some pipes to his truck when the white Chevrolet Lumina cruised up to the curb and Lieutenant Marsha Manning got out of the driver's side.

They'd had some good talks since he'd come, Scarborough and Mashkin, talks covering the breadth and length of the UFO phenomena. Sometimes the conversation had veered toward argument, but it was more debate than anything heated or ugly, which was what most of Scarborough's discussions with UFO afficianados had turned to over the years. Mostly, though, he'd slept, resting and relaxing and getting ready for what came next: his meeting with Ed Myers. But he'd come to respect and admire Mashkin, and this afternoon the man had to go out for a plumbing job. Naturally, it was best for Scarborough to just stay here, out of sight; but there was no reason he couldn't help the guy load pipe into his pickup truck, a voluntary action which Mashkin seemed to appreciate as much symbolically as practically.

"Hello, Marsha!" called Scarborough, after carefully setting the pipe down in the pile on the bed of the truck. "Glad you could stop by!"

"A trip to sunshine and warmth! I couldn't resist it."

"Walter, this is the young lady of whom I speak so often and well. I present, Lieutenant Marsha Manning of the United States Air Force, usually accompanied by a drum roll and dripping with medals of honor, but today travelling in mufti!"

Marsha snorted. "You're in fine fettle today, Everett! Hello, Mr. Mashkin."

"Please, Walter. Or Walt . . ."

"But never Wally. Walter's been taking good care of me, Marsha. I'm feeling sort of human again."

Mashkin winked. "I just been feedin' him some beer and some food and fertilizing his roots a bit with some jawin'—but come on in, Marsha. You come a long way, and maybe you could do with a nice long-neck bottle of beer yourself."

Marsha smiled, clearly charmed with Walter Mashkin. "Why thank you, Walter. But I thought the plane was landing in New Mexico, not the Smoky Mountains."

A grin split Walter's face. "Well, let's just say that I brung a bit of Tennessee to the Southwest, Marsha. And holdin' onto my accent like a miser holds onto his money! Now ain't you got a suitcase? I'll bring it in while you go in with your pal here and do some catchin' up!"

Marsha gave him the keys and pointed him toward the trunk with thanks, and then allowed herself to be ushered into the cool stucco house. For his own part, Scarborough was startled not only at how glad he was to see Marsha Manning, but at the effect she had on him emotionally and physically as she walked with him, the familiar scent of her Opium perfume and her shampoo dancing under his nose. Marsha wore good green culottes that accentuated her large, rounded hips, and a crisp white blouse that fairly bloomed with her large, high breasts. What's more, she seemed to have lost a few pounds since he'd last seen her—and she was wearing a little more makeup than usual, her long curly hair brushed out to kiss the light rose-colored linen jacket. The warmth in her eyes could not be mistaken—she was glad to see him, too, and clearly very happy that he was in decidedly better spirits than when last they met.

"You look very nice, Marsha," he said, getting the beer Mashkin had promised her from the refrigerator.

"Thanks. And you, sir, look like a plumber. An excellent disguise." She accepted the beer and sat down at the kitchen table.

Scarborough looked down at his apparel. Red-checked flannel shirt with sleeves rolled up. Worn Levi's jeans. Torn Keds sneakers. "What, you don't like my Air Jordans here?"

"I'm sure they'll put you in good stead at the Publishers' basketball match that decides our fate."

Scarborough grinned. "I am pretty good at dunking the ball." And at dribbling too, he thought, admiring her long legs as she stretched them out. Whoa, fella, he told himself. It's nice to know that the juices aren't all dried up, but there are lots more serious things than the visual delights of heterosexuality.

"Speaking of balls . . ." she said, pausing for a sip, letting the last word hang in the air. "I hope we haven't dropped ours here. It was hell getting to come out here, but I hope it's in time."

Scarborough sat down beside her and put a reassuring hand on her shoulder. "Things are going well. I'm almost positive I know where they're keeping Diane and Tim."

"Well, I had to come out here anyway to follow up on this business I dug up at Wright-Patterson . . . And I figure I might be able to help you as well, since we're taking about Air Force property here."

It was time for a frown. "Marsha, you've done so much for me. And I'm really glad you're here. But I just don't want to jeopardize your career . . . Christ, your very freedom . . . For me. I just can't ask for that."

"Well, I'm giving it, whether you like it our not—if I can." Her features hardened into determination. "What's the story on your CIA contact?"

"I'm to call his house tonight. His son David has been given the name of the place where I'm supposed to meet him tomorrow." Scarborough heaved a sigh. "Marsha, this is going to work. I know it. I'm going to get in for Diane tomorrow."

He looked at her seriously. "Yes... Air Force. Kirtland... Maybe you *can* help me."

"That's what I'm here for."

"No, in a way that you won't have to risk your life or anything else, for that matter."

He was interrupted by Mashkin coming through the front door and hauling two bags into the kitchen entrance and then letting them down with a gasp. "Whew. You women don't travel light, do you?"

"Gee, Walter. I can carry those bags just fine!"

"Big girl like you, I'm not surprised. Feels like these things are loaded with lead pipe!"

"Bullet-proof underwear!" she joked.

"Reckon I'll just set these down awhile and set my butt down, too. Teatime, right? I'll just get myself some tea." Mashkin got himself a beer, flipped off the top with the opener, leaned against the counter, and toasted his guests. "My, don't you two make a handsome couple!"

"If you consider Mutt and Jeff handsome!" said Marsha, laughing.

"So I hope I didn't interrupt nothing real intimate here."

"No, no, of course not, Walt. We were just discussing the plans for tomorrow. My meeting with Ed Myers..."

"And how I can help him at Kirtland."

"Yes, I thought that if I find Diane—and maybe even Tim—at that complex, and can get her into the lawful, legitimate part of the Air Force—well, then Marsha can make sure that she is protected."

"Well fuck me, good buddy, but won't that put your dick in the pencil sharpener? Pardon my vernacular, ma'am."

"It's Diane and Tim I'm worried about. I'm confident that if I get caught by the proper authorities, I can prove my innocence. But there is simply no way I can turn myself in and also find out what happened to Diane."

"I see. Well, I *guess* that makes some sense. So that's why Marsha's here, huh?"

"It's not the only reason," Marsha said.

Scarborough sighed and rolled his eyes. "Look, let's not start up on this, okay? I'm here to rescue my daughter and to try to get to the bottom of this operation, *not* to chase after little green men!''

Mashkin's eyes sparkled. ''Well, Marsha. You got my little pointy ears up! Would you care to fill in with some more information?''

''I see that *you've* got an open mind, Walter, unlike certain stubborn jerks who will not be named in present company.'' Her eyes smoldered toward Scarborough.

''Look, I'm not saying that there might not be something to what you found in that storage hangar, Marsha. I'm just saying that it's low priority now . . . It really has nothing to do with getting Diane out, or finding out what's going on at that complex.''

''You don't know that for sure, Everett. It could have a lot to do with what's going on. It's part of the whole cover-up, after all!''

''Please, please folks. You're killing me with suspense.''

Scarborough shook his head and threw up his hands in surrender. ''Lieutenant Manning here thinks she's found the Roswell crashed-saucer of 1947 at Wright-Patterson Air Force Base.''

Walter Mashkin's eyes grew wide and round. ''No *shit*!''

''Well, it was all very top secret on the computers . . . so I had to go take a look.''

''How the hell did you get in?''

Marsha Manning told the affable man the whole story. By the time she was finished, Mashkin had finished his beer, but had been too entranced to get another out of the icebox.

''Wow!'' he said when Marsha was finished. ''That's *terrific*! I been foolin' with this UFO business for twenty-five years, and all I got to show for it is a nice collection, some good friends and a sighting. You've stumbled upon actual *proof* that the saucers not only really exist, but that there's a cover-up.''

''We know there's a cover-up!'' said Scarborough impatiently. ''What we don't know is why, and I'm just not ready to start accepting that there are actual flying saucers from Xenon involved.''

Marsha looked at him with a sadness in her eyes. ''I really wish you'd believe me, Everett. This hurts.''

"Marsha, I *do* believe you. I just don't want to swallow whole the conclusions that you've jumped to!"

"Stalemate, then," Marsha said.

"Not really." Mashkin's face had a strange expression to it. "Ev, buddy. We got ourselves a free evening, right? I mean, after you call up the Ed Myers fellow's kid."

Scarborough shrugged. "Yes. I suppose so. Why?"

"Well, what would you say to a little drive in my pickup truck?"

"Whatever for?"

"Hmmm. Might be a bit crowded in old Miss Gertie—that's my pickup. Do you suppose, Marsha, we could use that nice new rental car? I'd pay for the extra mileage."

Marsha shrugged. "Unlimited mileage, Mr. Mashkin."

"All right, all right, so much for the suspense!" said Scarborough, irked, despite himself. "Walt, where the hell do you intend to take us?"

"Well, don't rightly know if I can take you anywheres. Gotta go and make a certain phone call first. Just checking on matters of feasibility and availability, that's all." He tipped an imaginary hat to Marsha. "Now, if you'll excuse me, I'll just retreat to my Den of Iniquity and Inquiry to make the necessary phone call."

Walter Mashkin left. Soon after came the sounds of an opening door and footsteps pounding down the cellar steps.

"Wall ya'll . . . do ya thaynk that's whereuh he keeps his alien moonshine!" drawled Marsha.

Scarborough laughed. The tension between them was immediately broken. "Actually, he's got a very nice basement. He's got a word processor down there which he uses to correspond with UFO magazines and other UFO afficianados."

"Quite a character!"

"You don't know a fraction of it!" Scarborough rolled his eyes, but he smiled. "I like him, though. What's the term? Salt of the earth?"

"Salt of the ethereal planes, more like!"

"No, he's more the hardcore, brass-tacks kind. Amateur UFO investigator. Probably very thorough, very logical—all revolving about a center a little off whack."

"I don't know, Ev. After all that's happened, all that I've

told you, you're still not willing to open up at least a teensy-weensy possibility that there may be something to it."

Scarborough sighed. "Let's not argue about it. I respect you deeply, and I'm sure that what you found in that hangar *is* very important. Let's not jump to any conclusions though, okay?"

"You're so cute when you're patronizing," Marsha responded tartly.

"Glad you think so." His features softened as he looked at her. God, she could be infuriating, but ultimately he rather *enjoyed* it. She was certainly her own person—she was like Diane in that. She had a brain in that pretty head, no question of that. A little batty at times, but nonetheless, it kept him on his toes.

"It *is* good to see you though, Ev, even though you can be the most infuriating son of a bitch this side of the Milky Way."

He had the sudden urge to reach out and hold her, but he checked it. "It's good to see you too, Marsha. I can't tell you . . ."

A sudden electrical charge hovered between them. It almost took Scarborough's breath away as he looked into the warm brown of her eyes, and although they were feet apart, it felt as though her hot, generously female body engulfed his own.

However, the sound of Walter Mashkin's work boots clopping back up the steps interrupted this intimate moment. Marsha turned away, her face slightly flushed. Scarborough had the terrible feeling that he was blushing. He went to the refrigerator for another beer, turning away.

"Folks, it's a go. We got ourselves an eight-thirty appointment with an acquaintance of mine tonight."

Scarborough pulled the beer out and made a show of opening it clumsily. "That sounds all right, Walt. But exactly who are we seeing, and *why*?"

A gleam appeared in Walter Mashkin's deep-set eyes and a smile curled the corners of his mouth. "Well, you see, Evvie, it's like this . . ."

CHAPTER TWENTY-SEVEN

The sun was gone, but the last rays of sunset still ridged the rugged landscape with soft shades of crimson and yellow light blueing skyward into darkness and stars. A desert breeze fluttered Marsha Manning's scarf as she got out of her Lumina and waited for Scarborough and Mashkin to disembark. Standing on the sandy road fronting the beaten old house on the outskirts of nowhere, his nose picking up the telltale smell of garbage and a poor sewage system, Dr. Everett Scarborough wondered why the hell he'd allowed himself to be dragged along for this fool's mission.

To humor Walter Mashkin, of course, was the immediate answer. The man, after all, had been a godsend these past days, and promised to be just as much help in the future. What harm, he'd told himself, would it be to go and visit some creaky and deranged UFO-nut friend and listen to a crackpot story? After all, it wasn't as though Scarborough had never done it before . . . In his research he'd collected enough crazies to start up a franchise system of loonies. He'd even considered a book once about all-American eccentrics called *MacNuts*. He supposed he could deal with one more on his evening off. Besides, it would take his mind away from the impending date with Ed Myers tomorrow. And it would prevent him from having to be with Marsha Manning, alone.

The house was an old and ramshackle adobe hacienda affair, scoured grey by the wind and cracked by the beating sun. Weeds and dead bushes poked up around the walls like dry, skeletal fingers from vegetable graves. The place was located well out past the sprouting Albequerque suburbs, but it harkened back to the old days of the Dons of the Spanish who had settled and once ruled this wild land.

"Wow. The Air Force really treats its own well," he said, raising an eyebrow at Mashkin.

"Oh, Jenkins got it cheap at a tax auction years ago. Really doesn't have much to put in it, though I suppose if he ever takes up a few of the offers to sell off some of this acreage to developers, he'd have plenty of money. Still, he likes his isolation. Probably keep this place like this until he kicks the bucket. But come on, my friends, he's waiting for us. It's not often that Jenkins lets people come visit. He'll probably even break out his tequila with the extra worms at the bottom." Mashkin winked at the cringing Marsha and then gestured the party forward.

The slender man with the Tennessee-accent led them down a pathway of broken flagstones, along which skittered the occasional lizard.

"I just hope this visit is worth it," said Marsha, hanging back away from the loping Mashkin, staring down in distaste at the reptiles and sorry state of the sprawling house's yard.

"You were the one who was so eager to come!"

"I know, I know." A moment of silence. "Did you get your message all right?"

Scarborough nodded. He'd gotten through to David Myers with no problem whatsoever. "I'm meeting Myers at a McDonald's just outside of Kirtland. He's got some doctored papers for me and Mashkin. We'll be the local plumbers, like I suggested. He'll get us both in, no problem."

"Good. We've already established where I'll be."

"Right. Backup. Officers' Club, right?"

The idea was that if—*no, dammit*, when—Scarborough found Diane—and Tim?—he'd hurry them over to Kirtland Central and there turn himself in. Once he knew that Diane was safe with trusted authorities as opposed to the CIA, he would allow himself to be arrested. Diane and Marsha could then attest that he was no where near MacKenzie's house when Mac had been killed.

She grinned. "I figured I might as well have a few drinks while I'm waiting for you."

"You sound like Camden."

"Who was supposed to show up, right?"

"Yes. Maybe there'll be a message on the phone machine

when we get back. Anyway, we can't wait for *him*, that's for sure."

"No. Camden will have to get the story third-person."

Scarborough shook his head. "I dunno. That guy has a spooky way of showing up unannounced."

"I just had a thought—you don't suppose that Camden could be involved with those people who shot your editor on the subway. . . who were following you in Baltimore?"

"No. I think I know Jake well enough by now. Take my word for it, Jake Camden works for nobody but himself."

"He seemed pretty out of it when he called me. Like I told you, that story of those guys that saved him really shook him up."

"Look, Marsha, let's just take things one step at a time. For right now, I'm just concentrating on getting my daughter freed. Okay."

Marsha executed a mock-salute. "Yes *sir*!"

Walter Mashkin reached the end of the gnarly path and stepped up onto a dirty stone porch. He knocked upon a wooden door as Scarborough and Manning hung back to give him and his friend some space. The knocks echoed hollowly through the big house.

"He *is* home, isn't he Walt?" said Scarborough, after a few minutes of knocking.

"Shee—it. Todd Jenkins hardly goes nowhere. Oh yes, he's home. We just gotta raise him, that's all!"

Walter Mashkin commenced knocking some more.

Suddenly, from the shadows past the porch, a click sounded: a rifle bolt-action.

Scarborough jumped despite himself. Manning grabbed his arm and squeezed.

"Can't you assholes read the goddamned sign!" barked a hoarse, raspy voice. "No soliciting! Now get your butts offa my land."

Mashkin stepped off the porch, his face breaking into a broad grin in the dying light. "Todd! Yo! We ain't tryin' to sell you nothin', you asshole! It's just me, Mash, right on time!"

"What, Mashkin! Walter!"

Scarborough noted with relief the rifle barrel lowering as a stooped, aging man with a mane of wild white hair strode into the light, staring at them through Coke-bottle glasses. "*Amigo!*

Shit, I forget you were hauling your sorry ass my way tonight.'' He stopped, casually and warily waving the rifle at Scarborough and Manning. ''So, who are these jokers?''

''Tarnation, Jenks. I *told* you. I got myself some guests. This here is Marsha Manning, and we brung along her pal—uhm . . . Doug Adams.'' It had been decided in the car that it was best to give Scarborough another identity for the time being.

Marsha lifted a hand to shake Jenkins's hand, but the man just stared at her like she was crazy. ''Oh yeah.''

''Well, you shithead, aren't you gonna ask us in! You and Marsha have got some stuff to talk about!''

''We do?''

''Yeah!'' said Mashkin, grinning over at Scarborough. ''The Roswell saucer crash! 1947!''

''You interested in hearing about that, Miss Manning?''

''I most certainly am.''

Jenkins nodded. ''In that case, I hope you got lots of room in your ears, because I'm going to fill them mighty full!''

The interior of the Spanish-style hacienda was large and drafty, except for the smaller rooms in the back where Todd Jenkins led them. The place smelled slightly off, in the way that bachelors' quarters tended to get, only this house had a peculiar desert quality, as though rattlesnakes or lizards lived off in the corners somewhere (all with definite whiskey habits). Against a wall were stacks and stacks of books, some in cardboard boxes, some not. Against another wall were piles of old magazines. Rounding out this ramshackle library were loads of old newspapers. Apparently, Todd Jenkins threw absolutely nothing out.

He seated them at a long cracked and warped table and then, at Mashkin's request, went off to get some beers.

''Todd Jenkins is a retired captain of the United States Air Force,'' said Mashkin, looking terribly pleased with himself and remarkably comfortable in this strange pigsty desert mansion.

''So you said before,'' Scarborough remarked, sitting uneasily in an ancient and creaky chair.

''You just wait, Ev. He's got a story that's gonna change your life!''

''We'll see.'' He turned to Marsha Manning, who to his

chagrin looked extremely intrigued. For his own part, Scarborough felt decidedly uneasy. He wished now that he had been selfish, had just said no to Walter Mashkin's wish that they meet this guy. He should have just stayed home and gone to bed early. Tomorrow was an absolutely vital day. What the hell was he doing spending the night before with a certified UFOol, ready to listen to some crack-brained, hoary old tale of aliens and flying saucers.

But even as these angry, knee-jerk thoughts rushed through his head, he caught himself up and told himself to cool it. This was the old Scarborough's attitude. He'd changed now, hadn't he? Well, somewhat, anyway. Enough to at least give some time to listen tonight. He owed at least that much to Mashkin. And if Marsha Manning wanted to be here, and wanted Scarborough here with her—yes, he owed that to her as well.

So sit still, you uptight asshole, he told the old Dr. Everett Scarborough; *shut up and listen awhile!*

When Jenkins came back he was carrying a six-pack of Dos Equis beer. He opened four, and passed three around dutifully, his mop of white hair flopping into his face as he did so.

Walter Mashkin toasted the air, and they all took a ceremonial sip.

"So then," Jenkins growled. "Walt here tells me, miss, that you've something real interesting on Wright-Patterson."

Marsha nodded and licked her lips tentatively. "Yes, sir. That's quite true." She took a nervous sip.

A moment of awkward silence hovered.

"Dammit, go ahead and tell Jenks what you told me, Marsha! He ain't gonna bite!"

Marsha cleared her throat and began. "You have to understand, Captain Jenkins, that I'm a specialist in computers. That's how I found out about the entry concerning the Roswell material. That's how I got a chance to go in and take a look at it."

Jenkins's eyes started glittering. His head started to nod like a dashboard doll. He seemed to be focusing on every word that Marsha Manning said, drinking it all in hungrily.

While Scarborough listened to Marsha's tale again, he sipped at his beer. He had to admit, this was pretty convincing stuff;

nonetheless, his mind worked overtime, out of habit, looking for a logical explanation.

When she was finished, Jenkins swept thick stubby fingers through his white mane of hair, opened another Dos Equis, took a long drink, clopped it down on the table, and sighed. "Those mother fuckers."

"Pardon me?" said Marsha clearly a little taken aback.

"Marsha," said Mashkin, grabbing himself the last of the beers. "You gotta understand. The United States Armed Forces put this man here through some real hell."

"War *is* hell, isn't it," piped up Scarborough.

"I'll take any kind of war, any day, other than what those assholes did to me!" growled Jenkins. He took a furious glug of beer and glared with red eyes at Scarborough. "And all because I was just trying to tell people the *truth*!" He turned a gentler gaze upon Marsha. "Ma'am, I've been waiting over forty years to hear those words from someone like you. Hell, they wouldn't let me within spitting distance of Wright-Patterson! And you—you just brought it all home to me like a balm from Gilead."

"See!" whooped Walter Mashkin. "What I tell ya, Jenks! You're vindicated. I can feel a whole 'nother article coming on about you, buddy!"

"Whoa there, just a minute," said Scarborough. "This doesn't prove anything."

"Give it a minute, Ev," said Mashkin. "I told you, it was Jenks's story I brought you out to hear. He just got all excited over hearing about Mashkin's little escapade, that's all. Ain't that right, Jenks."

"Excited? That's hardly the word. But I guess I do owe you my story now, so here goes."

The man scratched his crotch, drank some beer, and started.

"You have to understand, I was a career officer. Maybe you wouldn't understand, you not being in the service, sir—but I think the lady would. I had my name signed in blood even before the Nips bombed Pearl Harbor. I flew bombers in World War II in both theaters. I didn't know nothing else—and I sure can feel for those poor fuckers who come back from Nam with fried brains and their thumbs up their asses, cause once you get in the Force, it's damned hard to get out gracefully. Am I right, Lieutenant?''

Marsha Manning nodded glumly.

''Yeah. So I'm in for the long-term, you know. I'm stationed over at Roswell Army Base. It's 1947. Lot of my buddies went on to college with the GI Bill. Lot of 'em didn't make it out of WWII alive. Me, I was a lifer, and I went where they told me to go, and if they told me to fart, I asked how loud. So it's July 1947, and it's a boring day like all the rest, and suddenly a crew of us get some real strange orders. Seems as though something real strange buzzed the area the day before. Seems as though something big *crashed* about twenty, thirty miles away in some ranchers' range.

''Well, if you know anything about UFOs, you know that 1947 was the year that people started seeing the things a lot. Me, I figured it was the Hiroshima heebie-jeebies and didn't take much account about it. But we get these orders, like I say, to take a couple trucks over to this rancher's fields and have a look.''

''You were *there*,'' said Marsha excitedly. ''You were there at the Roswell crash.''

Scarborough listened attentively, but frankly, he had heard so many different stories about that event (most of them unprovable, some eventually exposed as outright *lies*) that he had to be doubtful. It had happened over forty years ago, after all—memory tended to start warping after about five minutes.

''That's right. Damnedest thing you ever saw. This rancher, name of Brazel, he was just standing there, not knowing *what* to do, and I don't blame him. There wasn't much *to* do, and so he just called in the sheriff who did all he could do, which was to call up the government, who sent in the military. Things were kinda crazy, and the military would sure have handled things different if they knew what they had on their hands. The top guys anyway. Sheesh, what a mess. Man, I took a look at what was there, and suddenly this great big black line came diving down into my life, making everything up to that time, *Before Roswell*, and everything that followed, *After Roswell*. B.R. and A.R. Kind of like B.C. and A.D. you know?''

''Jenks would you quit your goddamn hemming and hawing and get on with the gol-darned story!'' Mashkin said.

"Settle down, Mashkin. For Christ sakes, it's my story. I can tell it any which way I care to." He took a swig of beer and glared at Mashkin.

"That's quite all right, Mr. Jenkins," said Marsha Manning, obviously just as enraptured with the story as ever. "We're in no hurry."

Jenkins scratched some dead skin off his mottled red nose and then banged his bottle on the table. Beer foam spumed up and oozed onto the gouged wood. "A fucking flying saucer," he said, the remembered awe and wonders till etched indelibly on his face. "A crashed-down flying saucer, just ripped to shreds. And that ain't all, let me tell you." His mouth worked without words for a moment, as though he were having difficulty pushing them out. "Bodies. Must have been maybe five or six bodies, too."

"Alien bodies?" asked Marsha eagerly.

Jenkins just ignored her, seemingly so caught in the remembered vision that he was transported back to the scene he was describing in Roswell, 1947.

"There was wreckage just lying all over the place, and a long gout of dirt dug up. What was left of the thing was just this smashed-up machine with guts of machinery sticking out, like this great big thick dish cut in half."

"Just like I saw in that hangar!" exclaimed Marsha.

"Yeah, but you didn't see no bodies, did you?"

"No."

"The rumors and legends go that they've got these little bald bodies packed in formaldehyde." Jenkins shook his head. "Just ain't so. First of all, these bodies lying around—they weren't bald and they weren't little. They looked pretty much just like you and me, as best as I could make out. Of course, often as not they were pretty blown apart; but what's more," Jenkins paused and took in a deep breath, "even though they'd only died hours before, they were half-decomposed already . . . and going real fast. Whew! What a stench. You could almost see them melting, like that Poe story—what was it? Oh yeah. The 'Case of M. Valdemar.' Putrid gore just dripping like candle wax. Know what I mean?"

Scarborough blinked. "Can't really say that I do, Mr. Jenkins!"

Jenkins gave him a dirty look. "You know, you remind me a

hell of a lot of that skeptic-scientist asshole, Scuzzborough—or whatever his name is. But then I know Walt here, and I know that Walt would *never* have anything to do with slime like that joker!''

Walter Mashkin grinned over at Scarborough. ''Never!''

The scientist just shrugged and leaned back in his chair, letting the men get on with his story.

''So that would explain why there's no actual biological evidence of alien beings—they decomposed too quickly,'' Marsha Manning said.

''Yeah, and Mashkin has been harping on that in his UFO magazines for *years*! But does anybody listen to him? No way. They've all got their heads in a hole or in the sky—they don't think about fucking *logical* possibilities! So anyway, where was I? Oh yeah, so there we were staring at these rotting bodies. Well, by the time we got some kind of biological team out, they were gonesville. I mean, pond scum has more viscosity! I always figured maybe they—the aliens, I mean—fix it up like that, so's they won't get analyzed if they get caught or die or whatever. Anyway, none of them were in the saucer, they'd all been thrown out, so we just hauled the thing up onto a flatbed and gathered up the material and hauled it off, all the way to Wright Air Force Base in Texas.

''Now, by this time I'm recovering from the shock, and I'm thinking to myself, Jenkins, you're sitting on top of the most incredible occurrence of the century.''

''It hadn't occurred to you that this might be some experimental ship you saw?''

''What, and the men inside melting away like that? Course not. Besides, I didn't tell you, we were looking inside, and, like Miss Manning says, the material was like nothin' I ever saw come down from the sky—and human beings just don't decompose that fast! So anyway, I'm getting ready my statement to the press. I'm gonna be in *Time* magazine, I think to myself. Edward R. Murrow's gonna interview me! Maybe I would get to be in the picture—back then, I always thought I might like to be an actor. Anyway, so there I was, my head filled with all kinds of dreams on one hand, and honest shock on the other—I mean, it's not every day that a ship from some other planet takes a crashing dive into your life. There's a little trickle of talk about a 'downed object' or 'crashed plane' out in Roswell, and then a little flurry in the paper about it being a flying saucer. A photographer even gets a picture of some of my buddies holding up that weird thin foil-like metal. Supposedly, the officer in charge of the whole thing—a Lt. Colonel Marcel is about to make an imminent announcement.

"Then, suddenly, the curtain clamps down. Stonewall! The press can't get shit from the Air Force. Flying saucer? Pah! That wasn't no flying saucer! That wasn't any amazing vessel created on some other world far, far away. The stuff that those guys were holding up was just regular old earth stuff. What came down you see—this is what the said!—was one of their weather balloons."

Jenkins cleared his throat noisily, shaking his head woefully.

"Shit! I know what these two eyes saw, what this nose smelled! That thing in Brazel's field sure as hell wasn't no goddamned weather balloon. And I'm pissed off—about to go off to a newspaper and give my story. But some MPs stop me before I can go and they cart me over to Wright. And who should be waiting there to see me but Marcel himself.

" 'I have orders from the very highest authority in the land, Lt. Jenkins,' he tells me. 'And those orders say that what you saw is a secret that is highly classified, since it bears strong upon national security matters. You are hereby ordered, under penalty of treason, not to say a word to the press or anyone else, for that matter, about what you saw that day. Do you understand, Lieutenant?'

"Well, what was I supposed to say? Fuck you, buddy, I'm going out and make an announcement over CBS that there's a cover-up? Shit, no—like I say, I'm a lifer. This was my career. And the military back then, hell, they weren't the softies they are now. A couple of peeps from me, and my butt would have been cooling in Leavenworth.

"No. So I swallow the orders and my pride along with it, and I keep the old trap shut. I do my job. But over the years, I watch and I listen, and I can pretty much figure out what's going on. Project Sign, Project Grudge, and of course Project Blue Book. Shit, they're just distractions! CIA and Air Force cover-ups. I follow it all, and I don't say nothing about what I saw that day. Officially, that is. From this source . . . But I got a mind, and I can write. So I make sure that the word gets out to the underground movement of saucer watchers. I keep pseudonymous correspondence with some wheels in saucer clubs. I write articles for magazines. I make sure that the legend of the crashed Roswell saucer stays alive, and I also point stuff out in the underground—how Blue Book is a cover-up and all. I do this for years. Hell, I'm still doing it. I'm pretty sure the CIA knows by now, but it's such a part of the culture it really doesn't matter. It was over forty years ago and nobody really cares about what happened that long ago. But I do. It's like an obsession. I still have dreams about that thing—those bodies . . . And I can't help but wonder what kind of world this would be if the government didn't cover the whole thing up. If they put it all in black-and-white in the headlines of the world that, yes, there are beings from other planets visiting us. I can't help but wonder if this world wouldn't have somehow been a better one for knowing that we weren't alone, that there were other creatures in this universe than us petty bozos. I just can't help but wonder . . .''

His eyes grew unfocused as his voice trailed off to a whisper, sad, lonely, melancholic. Silence drifted down upon the assembled party. Even Scarborough felt a momentary frisson of something lost . . .

And something stirred in him.

Something deep.

For a moment, he felt a twinge of nausea. He felt out of control. He took a deep breath, closed his eyes—and slowly, he focused back into command. He didn't know what had caused this sudden anxiety attack—but as he opened his eyes, aware that the conversation had continued, he was grateful that none of the party seemed to have noticed his temporary problem at all.

''. . . just incredible,'' Marsha Manning was saying. ''I must say, Mr. Jenkins . . . you've been a brave man to keep this guerilla warfare going.''

Jenkins shrugged. ''If I was truly brave, I would have stood up right there and then.''

''Hell, in those times!'' Mashkin grimaced. ''When they thought there were commies hiding under every other rock and they had lynching parties out to get 'em. Damn, Jenks. Government would have strung you up faster than I can whistle Dixie! You done the right thing, boy. The right thing." He turned to the others and jabbed a thumb at their host. "This man's so-called guerilla work was what helped keep the movement alive, way back when there was absolutely nothin' goin' on UFO-wise. I could tell you marvelous stories!"

"Thanks, Walt." The man was unable to hide his pleasure at Mashkin's open admiration and approval. "It's been rough, I guess. A man's gotta have somethin' to live for, and I guess this has been the albatross around this joker's neck. Three wives couldn't take it—they all left me. Not at the same time, of course." A slight smile touched his lips. "Guess I'm a son of a bitch to live with. But you know, Miss Manning—hearing what you've told me has made all these years of work—hell, and I guess at times of self-doubt—worth it. I want to thank you. And I want you to know that if there's anything that I can do to help you in your search, or even if you just need a place to stay—you just come here and I'll do whatever I can."

"Thank you, Mr. Jenkins. I truly appreciate it."

"Well, tomorrow's a big day, and we got lots to do, Jenks," said Walter Mashkin, slapping his thighs and then getting up out of his creaky chair. "So, if you'll excuse us."

"Hell, anytime, Walt!" The old man ushered them out with a great deal more cordiality than with which he had received them, paying particular attention to be polite and well-mannered to Marsha, Scarborough noticed.

"Now, you guys are going to *have* to promise that you're going to keep old Lieutenant Jenkins here informed about what you find out. Damned if I want to die not knowing what the hell is going on, and where that saucer and those bodies came from."

"Oh, we will," said Marsha. She took his wrinkled old hand and shook it warmly.

On the way back, Marsha took hold of Scarborough's hand, squeezing it hard and looking up at the wash of stars across the dark velvet sky.

"Tomorrow," she said. "Tomorrow, Ev, we *could* find out."

Tomorrow, thought Scarborough. But he wasn't thinking about finding about flying saucers or about discovering any aliens.

He was thinking about Diane.
He was thinking about his daughter.
Tomorrow. . .

CHAPTER TWENTY-EIGHT

Jake Camden was drunk.

He sat now on a bar stool of the Golden Wheels Honkey Tonk, "The Finest in Local Country Music Talent and Ribs," relishing the rosy glow that permeated life, the universe, and everything, watching the bartender pour him another Old Grandad straight-up to complement the cold, cold draft of Olympia beer that glistened in its mug before him on the wooden bar. Up on the stage, the country music band was digging deep into some serious Buck Owens and Merle Haggard material; and while Jake Camden usually wasn't a big fan of country-western, he had to admit that it sounded mighty good while your butt was on good Western wood with good country alcohol ahumming in your veins and a couple of long toots of Colombian cocaine keeping you awake and sweetening the jangles of the gee-tars.

The process of Jake Camden's inebriation had been very simple.

With the money from Schroeder, Jake had checked into the local Holiday Inn and was about to call the number he had for his good buddy Ev Scarborough, when he had himself a little thought which went along the following lines:

Gee, I've been such a good boy for such a long time, and my nerves feel like they're been stretched out on the rack by the Spanish Inquisition. You know, ole Jake's been through some real hell, booming through the country, dry as a lawyer's eye, taut and scared. And here he is, where he's supposed to be, ready to be of service to his friends and compatriots . . . Now tell me, doesn't this fine man deserve just a teensy-weensy drink in this arid environment?

Now, Jake had seen, not a quarter-mile away, an inviting-looking establishment, the selfsame Golden Wheel where his butt had naturally gravitated, and he figured as long as he was stuck in New Mexico he might as well enjoy some of the local color.

Just a slow beer, maybe two. A little steak dinner with a baked potato and onion rings. Then his soul would be in proper shape to put that quarter in that pay phone and make that call.

Of course, when the blonde in the tight jeans and well-filled Western shirt had crossed his path, things had gotten just a little bit complicated.

The Golden Wheel had turned out to be a pleasant enough place, spacious and air-conditioned with a shellacked wood veneer that gave everything a brownish barry sheen that truly warmed the cockles of Camden's heart. According to a sign, the band playing tonight was The Spurs, but Camden had no intention of staying. He'd parked himself in a dark, cool booth, ordered himself that steak and beer with trimmings, and had every intention of finishing everything within thirty minutes and heading out, when the woman had wiggled her way past him and sat at the bar.

Camden drained the last of his Olympia, promptly forgot the rest of his onion rings and baked potato (he hadn't even *thought* about touching his broccoli), and moseyed his empty mug over to the bar, a couple of spaces away from the woman to order another.

By the fifth beer, he was great friends with the lady (her name was Edith, and she was buddies with the band and a word-processor at a legal firm in town), and had been persuaded to come back and meet "the boys" in a back room where they were tuning up for the night's performance. A few more beers were drunk, a vial of cocaine was produced, and somehow in the haze and excitement of the girl's shining and admiring eyes (he'd told her he was a famous reporter on assignment in New Mexico and she got all hot because she recognized his name because her grandma got the *National*

Intruder at the Winn Dixie she shopped at), he'd neglected to remember that he'd given up that soul-stinging white powder. Although it hadn't turned him into a raving lunatic, hungry for more, the coke *had* sobered him up enough to make him think that he could start drinking boilermakers.

Result: a drunk Jake Camden. He was now awaiting his country princess's imminent return from the ladies' room. Little blurry self-recriminations flickered on the periphery of his hampered awareness. At the back of his mind he had every good intention of putting a quarter in that public phone booth yonder by the coat-check room and calling the number of that Mashkin guy, to check in with Scarborough. But every time he was about to do it, something distracted him.

Like another drink.

Oh well, tomorrow would be fine. *Everything* was fine. Absolutely everything. Scarborough was just holed up anyway, planning something. No hurry whatsoever, Camden told himself. He had a night to blow out the jams and the devil take the hindmost. Without this little party, he'd probably be no good to Scarborough anyway.

"Hi!"

He realized suddenly that the band had stopped playing, and that, without the thumping bass and the high whine of the pedal steel guitar, he could actually hear properly. He swivelled on his bar stool and there, lo and behold, was his belusted, her blonde hair cascading silkenly off her lightly spangled jean jacket, her breasts looking about to pop out of her blouse. She was wearing White Linen, one of Jake's favorite perfumes, and for a moment he couldn't speak, the erotic power of the moment was so great.

"Hi!" he mumbled, blinking.

"Didn't think I was comin' back, huh? Well, I just had to make a little run in the back room there, talk to some friends, talk to a White Lady." She made appropriate sniffing sounds. "You wanna go back and do some more, Mr. Famous Reporter?"

"Well, normally yes, but maybe I've had just a little too much tonight." Jake couldn't believe those words were coming from his mouth, but he just rode with them to see what would happen.

What happened was that Edith pouted.

She pouted and heaved a breathy sigh, her chest expanding toward him almost popping a button, and the glimpse down the fleshy slopes of her breasts was enough to almost cause Jake Camden to fall off his seat. "Ah, and I thought that you were gonna be so much fun tonight!"

Jake's feeble conscience was effectively crushed. The old creature slinked up out from the depths of his libido, slithered into the driver's seat, and firmly planted its talons onto the wheel of control. "You know, I do like a little fun now and then." A little devilish twist quirked the corner of his mouth; a glitter took his eye. "I didn't think that a sweet and innocent young thing like you liked fun in these parts."

She chuckled throatily, "Oh, maybe one or twice in a blue moon!" She winked at him, running a long fingernail down his arm. Jake got goosebumps despite himself.

He reached over for his beer. "Well then, perhaps you and I should repair to the back room and meet that Lady you were talking about." He stopped for a moment, took a small sip, cocking his head at her. "Shame to share it, though, which we'll have to do back there with the band. You wouldn't happen to want to go on back to my room at the Holiday Inn and kick off our shoes, really relax."

Her eyes shone with enthusiasm. "Jake, I'd love to. But I wanna catch the next set. All the songs are dedicated to me! I helped pick them out."

"Oh." Jake hung his head, visions of burst buttons fading. "Well, I suppose true fun is in short supply in this God-forsaken land!" He made a mock-dramatic gesture with his hand. "Whatever did the pioneers do for relief from physical pressure! No wonder they shot each other up!"

"You silly boy. I'm not finished yet. There's a nice big janitor's closet off the other side of the dressing room. Nobody ever goes there about this time. We could go there and we wouldn't have to share anything." She patted the top pocket of her jean jacket, touching her right breast in the process. "And I got plenty."

Jake stared down at where she held her hand. "I should say so, dearie. So what are we waiting for?"

Carrying his beer along with him for possibly necessary

liquid fortification later, he followed Edith's provocatively swivelling backside into the smoky gloom.

The janitor's closet proved to be just that: a room filled with brooms, mops, and cleaners stacked on a shelf, along with paper towels for the rest rooms and other necessary oddments. It smelled of ammonia, dirty mopping water, and concentrated floor-wax, but as soon as Edith turned on the light and closed the door behind him, all that Jake Camden could smell was perfumed female flesh sweetened by the breath of his beer.

Oh, God, he thought to himself. Heaven is a broom closet!

"So, private enough for you, Jake?"

"Sure, babe. It's great."

Jake was in the throes of a dilemma. He didn't know which to go for first, the babe, or the coke. Edith decided for him. She took off her jacket, shook that great lustrous mane, bestirring the air with yet more musky aroma.

A little unsteadily, Jake reached for her.

And got a palm on his chest.

"Hold your horses, hotshot," she whispered, teasing promise soaking her voice, her eyes. "First things first."

She fished the new vial from her pocket, along with the same small mirror and Exacto knife she had used before. A rolled-up five-dollar bill later, Jake was suddenly standing straight, his brain pushing ignition buttons to nonexistent rocket motors.

'Yeow! I forgot how nice this stuff is!"

"Hmmm!" A bliss-out smile slapped across Edith's madeup face and her blue eyes rolled up slightly between eyeliner. "Yes. Ben gets it real pure."

"Ben?"

She put the impromptu straw down by the mirror on a barrel of industrial-strength Mr. Clean and began to unbutton that taut blouse. "Now, then, Jake. Do you think that sport-fucks are better on coke? I do. I think that it's the biggest rush in this world. Now get those old pants off and let's do a little experiment, huh?"

The blouse was suddenly off, revealing a black Frederick's of Hollywood lace bra above an hourglass waist.

Jake Camden gaped for only a moment, then his speeded-up hormones kicked him in the upper medulla and he proceeded to fumble at his pants buckle. His trousers halfway down, though, he realized that he'd forgotten to take off his shoes. He proceeded to do so. By the time he was finished, standing in shirt, jockey shorts, and anticipation, Edith had divested herself of every scrap and stood before him, an impatient hand on hip, clothed only in arousal. "Don't take anything else off, Jake," she said. "Come here. Now!"

He found his arms suddenly full of naked woman, blowing away whatever was left of his neurons. Her mouth worked against his with slippery urgency, and before he could even think about doing manly type things like mauling her breasts or grabbing her buttocks, her hand was down in his shorts checking out his state of readiness for the main event.

"Yow!" said Jake.

"Come on, lover. Come on." Pretty soon Jake was in the starter's circle all right and feeling his physical response, Edith lay down on the floor. "I'm hot for you. I'm wet and hot!"

She certainly was, thought Jake as he stared down at the woman's erotic contortions on the floor. He'd never quite seen anything like this before. That country-western, blues-in-my-beer stuff must kindle some interesting flames in the female id! His appreciation had most certainly risen, along with a certain item of his anatomy.

"I'm just an Okie from Muskogee!" he murmured to himself, pulling off his BVDs.

He had just freed one leg and was starting on the other when the door swung open. A big, red-faced man peered in, a scowl perched above broad shoulders like Zeus on top of Mount Olympus pounding out his thunderbolts.

"Oh, Jesus," said Edith growing limp. "It's my husband!"

Edith wasn't the only suddenly limp thing in the room. Jake Camden looked up with wobbly smile and said, "Hi there! You must be Ben. I don't suppose you're going to believe this, but we're just working out a new kind of encounter-therapy involving nude telepathy!"

"Read my *fist*, buddy," said the man stepping into the room.

All in all, Jake Camden really would have rather skipped over what happened next.

Picking himself out of the scattered garbage cans wasn't difficult. Keeping upright on two legs was the real problem.

Jake did his very best though, and after a few moments of effort, he succeeded. The coke had pretty much worn off though, giving him the wobblies. And his head and chest were still hurting from the pounding that he'd received from that slut's husband. Fortunately, the guy'd let him put his clothes back on between punches, before tossing him out into the trash area. Unfortunately, he'd grabbed a bag of fresh garbage and dumped it on Jake as a way of saying farewell.

Jake didn't care to imagine what the guy would have done to him if he'd come in forty-five seconds later.

The hotel room, he thought. Got to get back to the Holiday Inn. Everything will be okay if I can lock the door, take a shower, and fall into bed.

Jake thrust himself forward, one foot at a time, glad to be mobile. Then abruptly he realized that he didn't know which way he was going. Through bleary eyes he reconnoitered his situation. The garbage bins were behind him. In front of him reared the back walls of the joint where he'd met his misfortune. He did a little drunken calculating, and steered himself down the correct side, which served as the Golden Wheels parking lot. He'd walked here, but from which way? That was okay, he told himself. Once he got out on the highway, he'd give the avenue a good up-and-down and be able to spot the familiar green and white well-lit Holiday Inn sign.

It felt as though there were some poisonous-spined toad behind his forehead, swelling and expanding with every malignant breath it took. He was in bad shape and he knew it. All he wanted now was to crawl into his hole and heal for a while.

I fucked up, Scarborough, he found himself saying as though presenting a confession. *I blew it. But I'll make it up to you, I swear to God I will. Just as soon as I get some aspirin and some sleep.*

The two men stepped out from behind a large black Chrysler van, blocking Jake Camden's path to the highway.

"Pardon me, gentlemen," said Jake, surprised at how slurred his words were. "But I have to go home."

The men stepped closer and Jake realized that they were behaving in a decidedly stolid and menacing manner.

"Oh shit, guys, your buddy already beat the crap out of me and tossed me in the trash. What *more* do you want?"

"Whew, this guy smells like shit," said one of the men. "You sure this is the bozo?"

"Yes," said the other. "It's him."

It was then Jake realized that they were wearing suits and ties.

These guys weren't any urban cowboys!

"Hey. Who the fuck are you?" he said, stepping back and trying to swivel away and start a run.

The men grabbed an arm apiece.

"Come with us quietly, Jake Camden, and you won't be harmed further tonight."

"You guys can't take me! I got *mysterious protectors*!"

Of course, where had they been when that jealous husband had caught him with the stacked blonde in the janitor's closet just about to play plumber's helper.

"Right, Jake. And a snootful besides!" said one of the men, tightening his grip.

Alarm raced through him. Fear sobered him.

Jake struggled, trying to break free and run back into the bar. They wouldn't be able to get him there! All he had to do was kick up a ruckus and get the cops to take him. He'd rather wake up in a jail cell than wake up dead.

One of the men tackled him. Together, they hauled him back up. With professional ease, one clamped a hand over Jake's mouth, cutting off a cry for help. They dragged him toward the rear door of the van, which opened as though magically.

"Cut the fight, and you won't hurt any worse than you already do," one spoke harshly.

This only provoked another paroxysm of struggle from Jake, who suddenly realized he'd much rather be in the hands of an irate husband, or even pissed-off drug dealers than these boys.

Because suddenly, through his alcohol- and drug-fraught mind, Jake Camden realized who these men were.

The blow was quick and efficient, straight out of nowhere, stilling Camden and sending him off into an uncomfortable and lonely darkness.

CHAPTER TWENTY-NINE

He was in hell.

Pain enfolded him like a living sheet of fire and he burned, thrashing and wailing in the darkness.

Everett Scarborough was in hell, which he did not believe in.

"No," he cried, rigid in his torment, willing away the agony. "No! This is not happening!"

Just as suddenly as they had started, the flames whisked away, and he hurt no longer. He felt himself being whisked rapidly upwards, and fancied that he could feel G-forces, as though he were in a vehicle. But there was no metal or glass around him—he was uncovered by machinery, or even the comfort of clothing. He was naked—naked in the darkness.

Suddenly, stars faded on before him. A panorama of stars, a breathless canopy of brilliance, planets most dazzling in the forefront.

He looked down, and saw that below his feet burned the sun.

The flames, he thought to himself. The flames had not been hell. He had been in the sun, and now he was being somehow lifted out into space . . .

Alone . . .

Alone in the universe . . .

Cold and solitary, drifting in the sea of stars.

The feeling of abject terror, of agoraphobia like he had never before experienced closed around him, clamping into him like hooks, tearing him apart.

And he screamed . . . screamed with a sound that engulfed him like a closing fist, shutting out the sun, the stars, closing up everything.

. . . And awoke, lying in the darkness.

"Scarborough," said a voice in that darkness, and it was a soft voice, though not comforting. "Scarborough, do you know who you are?"

"No," he gasped. "I don't know. God help me, I don't know who I am."

Upon his threshold of awareness, memories pounded like summer rain on the tin roof of a shed. Memories he could hear, but could not hear, could not touch. They felt totally out of time, out of space, as though they belonged to someone else. And yet, he knew with a profound conviction that they were most explicitly his *memories. His and nobody else's.*

And they spoke to him, they hummed to him, they sang to him. Of different times and different places. Threads of images, flashes of texture. A man he did not recognize, leaning over him, doing something to his head. A hospital . . . Was he in a hospital? He felt as though he were in a hospital, and yet it was like no hospital he had ever been in before. Nor was it just a man who stood over him. There were more men, and at least one woman, and with sudden realization, he knew that he was naked beneath the sheet that draped him, and there were prickings and unclassifiable sensations in his head.

And, peripherally, he was aware of machinery.

Glinting, ticking, working . . .

"Everett Scarborough. Everett Scarborough, do you know know who you are?" said the soft voice, monotone.

"Yes, of course," said Scarborough, breaking out of his haze. "I'm Everett Scarborough!"

"No," said the voice firmly, and giving no quarter. "No, you're not."

A cold hand reached down and pulled the sheet from his body, as he was aware of eyes, cold eyes staring down at his nakedness . . . And he saw the cold, hovering machines bear down upon him flashing and shining with drills and dripping needles like the jaws of some hungry carnivore.

He screamed.

The scream woke him up, and he lay there in the spinning darkness of the bed, shaken and sweating.

A moon was peering through the slats of venetian blinds, spilling a pattern of light across the rumpled bedclothes. Even though a cool breeze whispered through the open window, whistling through those blinds quietly, he felt hot—hot and terribly disoriented. He lay there, clutching his pillow like a life preserver tossed to him in treacherous waters. It took him awhile, but slowly the pieces of memory reassembled themselves:

The last beer of the evening, frigid and tart.

The smiling face of Walter Mashkin, wishing them goodnight.

Marsha Manning's soft lips brushing his cheek and whispering, "Sleep well."

He was in New Mexico.

And tomorrow he would meet with Edward Myers who would take him to the edge of that Air Force base—to find his daughter, Diane.

His name was Everett Scarborough. *Doctor* Everett Scarborough. And it always had been, and it always would be.

"Scarborough," he whispered to himself like a mantra. "Everett Scarborough. Everett Scarborough . . ."

"Everett?"

His heart seemed to skip a beat. He tensed. A voice . . . A voice coming from beyond the closed door. Not a man's voice, but a woman's. Gentle, and yet clearly concerned.

"Ev, are you all right?"

The door opened a crack and in the moonlight, he could see a pair of eyes peering in.

"Marsha?" He meant to sound cool, collected, and confident, but his voice emerged as a frightened croak.

She came in, with a rustle of her nightclothes: a short silk chemise over panties. Her hair was undone, floating down over her shoulders like a dreamy waterfall, and the smell of perfume and woman and sleep advanced before her like an invisible mist, comforting and familiar, and yet promising far more.

"Ev. Ev, are you all right? I heard you cry out."

"Nightmare. Bad one." He sat up, still a little groggy and unfocused.

"I'm not surprised. I'm not surprised at all. You've been under a tremendous amount of tension." She sat down on the side of the bed and tenderly caressed his shoulders, his back: gentle message. "My goodness, you're a bundle of *knots*! Maybe we should have taken Walt's advice, hmm? If only to have someone near you . . ."

It took only a few seconds of her rubbing, her nearness, to relax him, if only a little. He sighed with gratification and release. "Maybe . . ."

When they'd gotten home, after their "nightcaps," i.e., another beer each, Walter Mashkin had announced, "Well, I guess it's time for ya'll to get yourselves to bed. Big day tomorrow. Gotta get your rest." He winked with a glitter in his eye. "I took the liberty of putting your bags in with the doc's, Marsha."

Marsha had obviously been taken aback by this announcement, but it was Scarborough who'd said something first. "Why did you do that, Walt? We're not that way . . . umm . . . We're just friends."

It was Mashkin's turn to be taken aback. His jaw dropped and he did a double take. "Bush my whackers! You're kiddin' me."

"That's right," said Marsha, smiling. "Just friends."

"Sorry, folks," said Mashkin after an awkward silence. "There was so much electricity sparkin' between you two, I just naturally figured you two were one hot item." He took a gulp of beer and shook his head, grinning. "I'm *real* embarrassed." He smiled at Marsha. "In that case, love, any chance of you parkin' your stuff by my cowboy boots! *I'm* not crazy, and I know a good lovin' woman when I see one!" The tone of his voice and the tilt of his head made immediate jest of his proposition.

Marsha had laughed. "I'm glad someone appreciates me. No, Walt. If you've got a couch, I guess it's mine."

"Oh, no, no. I've got another guest room. No problem."

Scarborough had been afraid that he'd blushed throughout all of this, but if he did, Walter Mashkin hadn't kidded him about it.

Now, though, as he sat up in his bed, feeling the residue of his nightmare reluctantly parting from him, he wished that Marsha *had* slept with him, if only so that he could have had someone to grab hold of in his throes of terror.

"You know, Ev. I'm here for you if you need me."

"I . . . I . . ." he said.

"Shhh. Dear Ev." Her voice was soft and sweet velvet. "You don't have to say anything. Just hold me if you want to. Just hold me . . ."

He opened his mouth to say no, to say he didn't really need

that, that he was a strong person, and he would weather this brief aberration, that it would never happen again, no, not to him. But what came out was a sighing croak instead, and suddenly he found himself reaching out for her, gasping her name.

Her flowery hair swept around him like a curtain of sighs, and her body was warm and alive and achingly soft as it folded him up in its comfort. Her lips found his and they touched with a slippery gentleness that elevated him into a safer, kinder place. He swept his arms around her and caressed her bare back, revelling in the way she was suddenly all abandon, her breasts mashing against his chest hardening with urgency, her breaths coming quicker. He lowered his mouth to her neck and fed upon its heat, and she arched her back and growled with pleasure.

"Love me, Everett," she said demandingly.

The silent music of erotic heartbeats took sway and the dance of sex began. Scarborough lost himself in the sudden fertile smell of this vibrant woman, lost himself in the eternal *now* of deliciously slow ecstasy. She stopped him a moment and then slipped off her top. Her full, curving breasts hung in the moonlight, large, and all his, and she offered them to him like fruit. He licked and savored them hungrily, and she sighed with gratification, softly murmuring encouragement.

"Enough," she said. "That's enough. I'm ready for you now, Everett. I'm ready."

She lay back, the glimmer of the moon in her eyes, licking her lips with expectation and tension.

He realized suddenly that he was ready himself, had been even as she had kissed him. Gently, he removed her lacy panties, sliding them down over her large hips and off her legs.

Then he stood and removed his own clothing. The T-shirt came off with ease, but the jockey shorts were a little more difficult. He lay down beside her, his erection stiff and constrained against her hip, and he gently caressed her inner thigh, kissing her arm, her breast, her neck, her mouth. As his hand drifted over her pubic area, he felt her heat and her wetness. Yes, she was ready. More than ready.

CHAPTER THIRTY

When he woke up, it crashed down upon him like doomsday. After a moment of lying there, engulfed by the enormity of it all, Jake Camden thought, Jesus Christ, help me. I have the King Kong hangover of all time.

Now, this was saying something. In his checkered career, Jake Camden had experienced his share of hangovers. During his college days, he quickly found that they came in different shapes, sizes, and colors, but that they were all alike in that they were not at all enjoyable experiences.

And this one was a real killer.

His head felt as though his heart had somehow gotten sucked in with his brain and was pounding away for all it was worth to get out, using a hammer in its efforts to escape. His body was one big throbbing ache, clothed in flesh and agony. His mouth felt like a desert cesspit. His abdomen felt as though a group of Boy Scouts were down there, practicing their knot tying.

When he got his breath back to think about more than the fact that he was in pain, Jake thought, Where the hell am I?

This was not easy to ascertain.

For one thing, although it wasn't exactly pitch-black all around him, it wasn't real bright either. Gloomy would be the appropriate term. Through the haze pounding around his eyeballs, Jake Camden could discern shadowy forms—blocky shapes. A room, then. It had to be some kind of room.

"Gosh," he groaned. "I'm a regular Sherlock Holmes."

It was when he moved that things got more interesting.

He couldn't.

Not appreciably, anyway. He could wiggle his toes (bare, it

would seem) and move his hands, but there were restraining belts tied around his ankles, thighs, abdomen, and chest. He could move his head around, but that was just the thing he didn't *want* to move, it hurt so.

"Well," he whispered sadly to himself, after taking a few deep breaths to prevent a total nervous breakdown. "This is a another fine mess you've gotten us into, Stanley."

He remembered now, of course. Remembered all too clearly that fatal glass of beer with dinner, the rounds rolling in afterward, the blonde, the band, the irate husband.

And then, crashing down on him like those fabled hundred bottles of beer on the wall, came the memory of those men in suits who had swarmed in on him at the parking lot.

If he hadn't fully realized it at that point, what with his preoccupation with his present frail state of health, he certainly realized it now:

He was not only in trouble, he was in *big* trouble. *Deep* trouble.

He moaned softly to himself, as though the sound of his own voice might soothe him.

Jesus, they've got me tied down and what the hell are they going to do to me *now*?

He didn't have to wait long to find out.

As though aware of his return to consciousness, the room began to change. Lights crept up from corners like monochrome false dawns.

"Hey!" Jake cried. "What the hell's going on here! Stop fucking with my head. It's fucked enough. Who are you! What do you want?"

No answer.

The lights stopped, but they outlined the room sufficiently for Jake to realize what strange geometric configurations comprised it. The place sure as hell didn't look like anywhere he'd been before . . . The ceiling described an odd pattern, juxtaposed by what Jake took to be some kind of symbols—hieroglyphics—stitched here and there like wall tattoos. If he wasn't basically a cynic at heart, he'd be inclined to think that he was someplace incredibly silly, like . . . hee . . . hee . . . an extraterrestrial—ho-ho—spaceship—or something equally unlikely like that.

Of course, it *did* look awfully *alien*.

Maybe it was a dream . . . yeah . . . that was it . . . some kind of drugged-out nightmare. That blonde had spiked his drink with LSD or peyote or some other hallucinogenic and now he was undergoing some kind of visionary nightmare.

Of course, he'd never been hung over in a dream before, so that was kind of unlikely . . .

As though to immediately disabuse him of the nightmare theory, a door opened and something very real, very strange—and yet curiously familiar—walked in.

He totally forgot his hangover.

It was short, this thing, wearing some kind of silvery jumpsuit. It had almond eyes, a bald head, and a pointy chin and no nose; and although it looked something like the creatures that had been featured in Maximillian Schroeder's first movie, there was something ineffably so *different* about it that Jake's breath was taken away.

The creature walked toward him silently, and stopped in front of him. It raised an arm, and Jake Camden could see that held in its slender fingers was a wand of some sort.

The end of the wand suddenly lit with an almost magical, lambent glow.

The creature turned its moist eyes upon him, held the lit wand up, and pulled it through the air a foot above the speechless Jake Camden, moving its mouth, as though muttering an incantation.

Camden opened his mouth to say something smart like, "Please don't stick that thing up my ass!" or "Where have you been all my life!"

What came out was a scream.

It came out full throttle, and it seemed to last forever, tearing out his tonsils, his uvula, his teeth, and his tongue in its ferocious torrent. It felt as though there were some kind of meaty thumb pressed down on his scream-button in the back of his mind. Only halfway through did he realize that yes, for some reason, his whole being seemed permeated with a deep, all-encompassing fear beyond all explanation. He saw no reason to be afraid of this creature before him—not logically, anyway. He almost felt a sense of déjà vu, as though he'd been here before. He'd heard enough testimony about this particular scenario, written about it enough, read about it, even seen it on the big screen enough to make it passé, a cliche. And yet Camden, now smack-dab in the midst of its reality, felt as though the stops of all his adrenal glands had been torn out, and his blood had turned to pure adrenaline.

Just when he wondered if he was going to die from screaming, Jake stopped.

The creature had stepped back a bit, looking down at him with detachment and curiosity with its big, limpid eyes, as though not surprised but a little nonplussed at this reaction.

It waved the wand over Jake, and he could feel the fear subside somewhat.

With time to recover, Jake took in some hasty breaths and looked closely at the thing, straight up into its face. With his being not being ravaged by the fear, he was able to think a little straighter, a little better. From the time this supposed ET had walked in the door, he'd felt something was off about it, and now that it was up very close and he could look into its eyes, he saw that even though they were wet and glistening, there was something very wrong about them.

They didn't seem awake. They didn't seem *alive*.

The only thing that Jake could move was his mouth. They really should have put a restraint around that as well: It was his most powerful weapon.

"Look, you assholes," he yelled. "You may be able to fool farm hicks in Oshkosh, Wisconsin, with this bullshit, but you can't fool Jake Camden."

The creature stopped moving, and bent its head, peering at him curiously from a different angle.

"Come on, you turkeys can do better than this! What is this, Disneyland or something. You know you've got me at an angle where I really can't see, but dollars-to-donuts there are tracks and wire running out of the bottom of this thing. And don't think I don't know the word for it. Simulacrum! Am I right? And you've got me loaded down with exotic drugs, I bet. Well, thanks a whole hell of a lot, but what I really need is an aspirin, so if you give me that and a glass of water, maybe when I get out of this and write my book I'll be a little bit merciful."

Silence. The room seemed to resonate with it.

Jake was getting annoyed.

"You think I don't know what you've been up to?" Actually, he didn't, not really, but he was following the theory that he and Scarborough had cooked up after visiting that CIA farm in Iowa, and after Scarborough had visited that lady in Baltimore. "Okay, I'll spell it out for you. You're trying to make me believe that I've been captured and am being examined by aliens. Only its a sham, a mock-up, an elaborate form of disinformation. To cover what up, I haven't got the faintest idea—but that's what it is, isn't it? We've got some kind of ultrasecret government program here, don't we? Those guys who picked me up were wearing the standard uniform, folks. They weren't aliens and they weren't Men in Black. They were FBI or CIA or God-knows-what other fascist branch of this corrupt government—but that's what they were!"

Silence again. Jake took in a breath, but then held his tongue.

Was that a whir? A click? The shuffling and whooshing of hydraulic equipment?

The wand in the creature's hand dulled. The creature's arm fell to its side. It turned and left the room.

"So show yourselves, you bastards!" said Jake Camden. "Come out and face me!"

He felt a slight prickling at the base of his neck, as though invisible fingers had touched him and sparked onto his skin.

A relaxing wave of darkness fluttered down over him like a cowl, and he passed out.

When Jake Camden awoke again, he did not hurt so badly.

But he was still restrained upon a table like a specimen in some laboratory about to be dissected. Camden felt as though the worst of the hangover had passed—and yet, he was immediately aware of a *wrongness* inside his body, a wrongness that he immediately diagnosed as drug-related. Now, Jake Camden had done his share of drugs in his life, all the way from steady pot and LSD use in college, to experiments with Ecstasy and freebasing cocaine within the last few years. The cocaine had been his ultimate downfall (he didn't really count his drinking; he considered alcohol part of his job), but he'd handled that problem pretty much (except for that brief backslide last night). The psychedelic drugs he'd stopped using long ago. What was going on now in his head and body had a faint whiff of the psychotropic. Little flashes of disorientation and paranoia alternated with faintly altered perception. There were subtler elements. Stuff he couldn't get a handle on. Whatever it was, he knew that he'd been drugged, but he also knew that he somehow was still in control.

He was no longer in the souped-up, bizarre room. No, now he was in some kind of doctor's office. There were tables and cabinets: white and efficient-looking. The smell of antiseptics hung in the air, that familiar scent of the physician. He noted the faint gleam of instruments here and there. Stainless steel. He sensed the hypodermics, hidden away like wasps, their stingers behind whitewashed walls. Jake Camden cringed, the paranoia blooming in him again. God, he hated needles. Always had, always would. That was why he'd never done heroin, ever. Never been tempted. It would be too much like back when his old man had dragged him to the doctor's office for that interminable battery of unsympathetic shots in the arm or behind.

Wonderful, he thought. *And doubtless these monsters know that fear—it's probably written out and underlined on whatever report they have on me*.

He still couldn't move. That troubled him as well. However, again he had free use of his mouth. This time, though, he bided his time. He kept quiet.

He didn't have to wait long. Within a few minutes, the door opened. No golden-eyes extraterrestrial this time. No, this time it was that other obsession of Jake Camden's: a blonde. A pretty good-looking one too, he was happy to see—and he felt relief that he still noticed such things.

"That's all right," the woman said stiffly to someone outside. "He's restrained and under sedation. I'd prefer to deal with this myself, alone. If I have any problems, I'll hit the button."

She closed the door behind her and looked over to Jake, a strange mixture of concern, preoccupation, and eagerness in her eyes. She was dressed in a long white lab coat. Her hair was drawn back into a severe ponytail. She was wearing glasses and a frown.

On the other side of the room was a desk. She strode to this and professionally paged through an open book filled partially with print and partially with scribbled notes.

"Hmmm," she said and coughed faintly into a hand. "Well now, Mr. Camden. It would appear that you are quite the character. How do you feel?"

"Drugged."

"Yes." She held up a small list. "Here we have the exact drugs we've been using in this particular case, including the amounts administered. You have a remarkable threshold, Jake. I suspect, however, that should we do a complete physical examination we'd find a great deal of damage. What's the phrase? Burning the candle at both ends? Apparently, Jake, you've been using a three-wick candle. Also, it would seem that you are remarkably unreceptive to the scenario we created for you. Oh well, it doesn't work for everyone. Only for the highly suggestive. Now then, though. To work. What exactly are you after, Jake Camden? And why are you aiding and abetting a wanted criminal, almost certainly guilty of betraying the interests of the national security of the United States?"

"You mean Doc Scarborough? Shit, he's not betraying anybody. You guys are CIA, aren't you?"

"I'm the one asking the questions, Camden. Let me advise you to answer them plainly and truthfully. There are other methods of extracting information besides relatively painless drug combinations."

"You *are* CIA. Whew. The stories are true, then. Okay, shit, I don't want to get *tortured*, for God's sake. Don't your notes on me say I'm a physical coward?"

"Not precisely. I don't think you've ever been quite in this sort of situation before, have you?"

"Not quite. But I'm not a great fan of pain."

The woman got up and went over to him, leaning down close as though for emphasis or intimidation purposes.

"You'll tell us what you can about Everett Scarborough, then?'

"Sure. I guess. I really don't know all that much."

"You seemed to know quite a bit in our little scenario enactment about an hour ago, it would seem. This troubles me a great deal, particularly since you seem to have drawn some ugly conclusions. The implications are that you've done research that we are not aware of . . . And perhaps told others of your findings. Or worse, written them down." Her teeth clenched and Jake could smell a touch of perfume about her: a medicinal smell. "Is that it, Jake Camden. Are these findings sitting at some newspaper or book publisher, waiting to be published?"

"No! Hell no! We haven't gotten that far yet . . . I mean, I haven't . . . I mean . . . Well, there *was* this article about Scarborough. But it was mostly how he'd been framed. There wasn't anything about—umm—scenarios. Or alien simulacrums or disinformation or none of that stuff. Mostly because, sister, I haven't got the foggiest notion of why you're carrying on this loony nonsense!"

The woman relaxed somewhat and stepped back, folding her arms across her chest. "You don't expect me to *tell* you, do you?"

"No. I expect you to let me out of here! I'm an American citizen, goddammit. I have rights!" The old newspaper-reporter-in-a-jam speech—at least it gave him something to say, and the anger involved helped to clear his head. "I demand at least a phone call to my lawyer!"

The woman permitted herself a smile. "Surely you jest. You gave up any rights when you poked your nose into these matters of national security."

"National security! What bullshit! And that doesn't make any difference! Am I arrested? In that case, I get to call my lawyer."

The woman seemed to think this even funnier. "Mr. Camden . . . Jake. You are *not* arrested. You may leave whenever you want."

"Then take these straps offa me."

"Oh yes. One little item. We have to do a little unofficial interrogation."

"And scramble my brains up a little bit too, huh? Try to get me to buy this alien-abduction line? See if it will work on old Jake. Well, it *didn't* work. And you don't have to interrogate me, because I'll tell you everything I know."

"Will you, now?"

"Yes. And the first thing I'll tell you is that I've figured out who *you* are, lady."

"Oh yes?"

"You bet. You're Dr. Julia Cunningham—a regular CIA Dr. Mengele, according to Scarborough."

"Am I, now?" Although she tried to keep her composure, Camden could tell that he'd hit his mark. "You seem very sure of that."

"Well, you're blonde, in you're thirties, attractive, a regular ice queen—and you're an expert in psychoactive drugs. And don't you know that I feel a few of those blooming in my brain right now."

"Does that make a difference—who I am?"

"No, but it proves to me you're CIA."

"I thought you'd already established that . . . in your own mind, anyway."

"It proves we're on the right track. In the investigations."

"Again, I would think that you would be fairly certain of that after your recent experiences, Mr. Camden."

"The thing I can't figure, I mean, the big question I have . . . is *why*?"

"I thought that *I* was the interrogator, Mr. Camden."

"Oh, you are, you are. I just wanted to get a few facts straight."

"I think I've had just about enough of this," she said. She strode over to a cabinet, from which she drew out a pre-prepared tray of drug ampules and hypodermics. "Where is Dr. Everett Scarborough, Mr. Camden?"

"Ooops. That's the one thing I can't tell you."

"I thought not." She examined the bottles, then removed the protective guard from a disposable hypo. "Are you familiar with the case of William Buckley in Beirut, Mr. Camden?"

"Oh sure, the CIA spook that terrorist organization kidnapped."

"There was a doctor involved. A man who administered the drugs to William Buckley. Dr. Aziz al Abub. Are you at all familiar with the videotapes that were sent back to the CIA?"

Although Jake Camden did not report on the story, he was a news junkie, so he well knew the story of William Buckley. He cringed inside at the memory, the fear and paranoia making a return visit to his spine, running up and down with needle-sharp cleets.

Buckley was the head operative for the CIA in Beirut during the times of its greatest travail. In 1984, Buckley was kidnapped by a radical Moslem group known as *Hizballah*. Soon, the CIA began to receive videotapes of the progress made upon William Buckley. For although Buckley was a highly trained operative, fully prepared for what would happen to him in the eventuality of a hostage situation, the *Hizballah* brought in an infamous and quite amoral doctor who subjected him to a battery of drugs that practically wiped out his personality. By the time the fourth or fifth tape came, depicting scenes of outrageous torture, Buckley clearly had not only spilled his guts to the *Hizballah*, disgorging a treasure trove of classified secrets, but he was calling for both the withdrawal of American troops from Beirut and the end of American intercession in Lebanon.

William Buckley had not merely been brainwashed. His grey matter had been hung out on a line to bake and dry, and then crushed to fit the mold his inhuman captors desired.

"Yes," said Camden, suddenly feeling meek. "I'm familiar with those tapes."

"Good. Because you should know that as a doctor whose speciality is brain chemistry and behavioral modification, among other pertinent subjects, we are not only fully familiar with that very effective methodology employed by Dr. al Abub. We have our own methods, far more effective . . . And sometimes not quite as subtle.

"You see, Mr. Camden. The mysteries of the mind are slowly being unlocked. And I am one of those privileged to fully be able to delve into such matters unchecked, due to my placement in this particular system. I have discovered wonderful and terrible things . . . but the search continues."

A kind of glittery self-absorption took hold of Dr. Julia Cunningham's eyes, as though she'd just received an injection herself and was feeling a rush. She examined the needle, squeezing out the air bubbles in a tiny jet of fluid. She turned to Camden.

"I have the distinct feeling that you are going to prove to be a most interesting and worthwhile subject, Jake Camden."

As she stepped toward him, the hypodermic held out before her like the stinger in the tail of a black-widow spider, Jake Camden looked at her face, and recognized what he saw there, the look that usually made him run far, far away from a woman.

The look of sadistic pleasure.

Now, of course, he couldn't run anywhere.

Jake had the distinct feeling that this was more than just a career or a challenge of curiosity into the workings of a human being's mind.

Jake Camden had the distinct feeling that Dr. Julia Cunningham was *enjoying* this.

Scarborough came down from above her, craning down with the smoothness of intuitive motion, and slid inside her with astonishing ease . . . Ease and *rightness*, as though their angles were part of a greater whole, come together for the first time.

When he was fully inside her, she lifted her head back and crooned a soft murmur. "Yes, Everett. Yes."

He positioned his hands to either side of her and began the timeless and eternal stroking, heaven's slide, her aroma enveloping him in sensory ecstasy.

"Yes, Everett. Fuck me. Fuck me!"

It took him with sudden surprise. One instant he was in control, the next her velvet grip tightened in his most vulnerable spot, triggering inevitability. The weeks of fear and panic, of uncertainty and tension, all collected in one mighty heave, and rushed out in a trembling, shaking explosion of release. Quivering, he drained it all into her with a groan of pleasure, abandon, and disappointment. It seemed to go on forever, this liquid rush, and for a moment his consciousness threatened to fade away.

Cherry blossoms!

He awoke from a half-faint, blinking at her side as she stroked his hot forehead, whispering soft nothings.

Cherry blossoms?

Why had he thought of cherry blossoms?

And then, suddenly he was Dr. Everett Scarborough again—lying with a woman whom he knew he had disappointed. The phrase "premature ejaculation" boomed at the back of his mind like an accusation.

"Sorry," he said.

"Sorry? Whatever for?"

"Well, I didn't exactly last."

"Don't ruin it. It was lovely. It was beautiful. Please don't ruin it."

"I usually last more than a few seconds."

"What do you think this is, Ev?" There was scorn and amusement and caring all mixed up in her voice. "An Olympic event?"

"No. I just wanted to give you something more than a couple grunting seconds."

"Ev, Ev, Ev." She softly kissed his forehead and he felt his tension loosening again as he lay between her breasts. "You poor, sick little boy."

"Oh. Thanks loads."

"Don't be silly. You gave me yourself. You really did. I didn't want a stud, I wanted *you*."

"Well, I *can* be a stud. I really can."

"The male ego always astounds me. So, are you feeling better?"

"I guess so."

"Don't pout so. It doesn't become you."

"Maybe you were just too much of a woman for me."

"Yes. That must be it. Like no other woman you've ever had before."

"Hmm. Well, there was this divorcée in Cleveland once who—" He was cut off by the pillow slamming into his face.

They both laughed, and in a way it was as much a release as what had happened just before.

When they came up for air, they were in each other's arms again, quietly whispering, lovers again.

"Really. That was incredible. For me, anyway," he said.

"It was very special for me, too."

"God. I hate that word. Sounds like a sale at K-Mart!"

"Attention, K-Mart shoppers! Dr. Everett Scarborough has just had an orgasm!"

"And Lieutenant Marsha Manning hasn't!"

"I felt a little tingle or two."

"Oh thanks. More praise for the performance."

"Really, you worry too much. Have you ever thought that a gal might take such a short, passionate, and explosive reaction to her charms as a compliment?"

"Hmmm. No, I guess not."

"Well, I did."

"You are a sweetheart."

"Ego soothed?"

"Healed. Redeemed. Resurrected on the third day and ascended into heaven."

"Good. Let me show you something then."

She kissed him firmly and soundly on the lips, and then slowly traced the outline of his jaw with her tongue. She nibbled expertly at his ear, and he began to stroke her back in the way of returning the favor.

"No," she said, rearing back and shaking her hair, her breasts pendulous yet firm before him. "I don't want you to move. I don't want you to stir a muscle unless you can't help it. This is for you, Ev. All for you." She chuckled throatily, and then buried her mouth in his chest hair. She tongued his nipples hard, and then dreamily drifted down his abdomen, her hair tickling his skin, expert fingers performing little curlicue dances of delight.

This tactile sensory cascade comingled with the smell of her, the fierce and abandoned and yet intelligent sexuality that had been implicit and yet so far a subtle part of Marsha Manning, was almost more than Everett Scarborough could stand. He drew in air, expulsed it with a gasp. With perfect timing she descended to his thighs and began licking upwards.

Scarborough suddenly realized what she intended to try to do. "Marsha," he said. "Really, I think I need more time. I mean, ego aside, I am fifty years old!"

"Shh! I want you to be quiet, Scarborough. Just relax. Enjoy. I'll show you . . ."

He let his head fall back into the pillow, a little doubtful, but willing to go along for the try. God knew, the effort would certainly take his mind off things like nightmares, suspense, and tension—at least of the dreadful, nonerotic sort.

Marsha crept up his legs slyly and knowingly, and then leaned down and kissed his genitals, as though in a friendly, how-do-you-do kind of way. She let her hair gently drift over his crotch, drawing sweet and subtle sensations, and kissed the perimeter as though mapping out her territory. After long, drawn-out moments of delicious and surprising touches, her lips lowered and gently took his glans between them, teasing it with her tongue.

Then she took the entirety of his limp shaft in her mouth and grew totally still.

Scarborough kept waiting for her to *do* something. Suck, stroke, kiss, anything.

But she stayed totally still, and he suddenly realized that he was holding his breath. Her mouth was around him, and he could feel its firmness and the wet of the saliva, but nothing more.

Then, so slightly that he hardly noticed, she began to move. With infinite care and patience, she started a soft but insistent back-and-forth motion, her wet mouth making it almost frictionless—but not quite. Sensation began to bloom in Scarborough's groin. A deep, throbbing need, drawn from the well of his soul like cool dark water, bubbling up toward this expert and loving mouth.

She drew him out, lifting her head back in exultation as she detected the change.

With great surprise and more than a little pleasure, Scarborough realized that his penis was slowly growing in her mouth. She mumbled praise and amusement as she started a stronger stroke, beginning to massage his testicles in expert counterpoint.

"Jesus," he said. "I don't believe it."

Marsha giggled.

"Better use it while I've got it."

She came up for air. "Quiet, dear. I'm enjoying myself."

She licked him playfully and then proceeded to pleasure the hell out of him.

Just before he could bear it no longer, just before he felt as though all of his insides were going to come blasting out, she stopped.

"Okay. Okay, Ev. I think we can assume you're hard enough now."

She straddled him and slipped her sex down on him like a fleshy glove. She gasped and groaned as he entered her and then began to move up and down, taking her own pleasure.

For his own part, Scarborough was feeling rather befuddled and benumbed—and he realized that he was at the point where most of the sensation had been milked from his nerves. Therefore, he could take his attention away from his own concern of a premature end to this startling session and get lost in Marsha's enjoyment.

She moved on him as though his penis was her own private plaything, letting it stroke the exact portions of her interior that pleased her, teasing herself and then thrusting hard and harder, her breasts jiggling before him, growing sweaty and taut. Her intensity was amazing as she worked her way into a wild frenzy, desperately reaching up and up for fulfillment.

When it came, it was like a thunderous breaking wave.

"Everett!" she gasped, arching her back so far that he thought she was going to fall away from him.

Just as suddenly, she was back on top of him, her crotch trying to merge into his. She bucked and trembled, she groaned and she spasmed as though a high voltage wire was rammed inside her and not just a rod of flesh. She quivered and shook for what seemed an endless time, and when she was through, Everett Scarborough realized that he had come, too; only his orgasm had been totally lost and consumed by the power and dominance of hers.

She collapsed against him, perspiring, breaths coming jaggedly. They lay together in a pool of scent and cooling heat and release, quietly revelling in the afterglow.

"Wow," she whispered, nuzzling her nose against his arm.

"Wow," he returned.

"I can't stop thinking about what Walter said. About electricity."

"Hmmm. You know, Mac said that too. I just thought you rubbed me the wrong way."

"Just think. We could have been rubbing each other all that time . . ."

"No. This was the right time, Marsha."

"I know."

"I had to reach out for you. I had to . . . realize that I'm not alone. And you showed me. Thank you."

"Welcome, Dr. Scarborough," she cooed softly into his ear. "You're very, very welcome. And by the way, if in the morning you run across any pieces of my brain lying about—on the floor, between the sheets—would you please return them? I need all of them."

"How will I know they're yours and not mine?"

"They'll have 'Everett Scarborough is the most *wonderful* man!' written all over them."

"Like I said, how will I know they're yours and not mine."

She punched him lightly on the arm. "You silly creature."

He smiled and drew her close, and together they drifted off into a safe, nightmareless sleep.

CHAPTER THIRTY-ONE

Everett Scarborough sat in the cushioned plastic booth, leaning over his cooling cup of coffee and his untouched Egg McMuffin, wondering why he'd allowed Ed Myers to set the meeting place at a McDonald's Restaurant, of all places. Scarborough abhorred McDonald's and all its hamburgerized ilk, with their teeth-rotting treacle-greased garbage food. Usually, he wouldn't be caught dead in such a place. But Myers was right, of course. It would be easy to find; it would be inconspicuous because of its banality. And, of course, it was just a couple of miles from the gates at Kirtland Air Force Base.

Most of the breakfast traffic was gone now, and a young man swabbed the floor with a mop, while an old lady put a fresh tray in a cash register. The two extreme levels of minimum wage, thought Scarborough, both happy with the extra cash while Roy Kroc's corporation sucked in the big numbers, preying on the gustatory weakness of children and the cramped time of adults.

Scarborough sipped at his coffee and cringed. Awful. He looked at his watch.

Nine-twenty.

Myers had said he'd be there between nine-fifteen and

nine-thirty, which meant that he had another ten minutes to go. But for some reason, Scarborough had a sense of ill-ease sitting on his stomach, preventing him from eating the congealing breakfast or drinking much more of the turgid coffee.

"Yep," a voice hollered in his ear. "I'll say one thing about these places. They keep their urinals clean enough to piddle in!"

An overalled body plumped down across from him: Walter Mashkin in his grubby work gear. He attacked his Big Breakfast of pancakes, eggs, bacon, sausage, and biscuit with relish.

Scarborough really couldn't complain. He wore almost exactly the same outfit. Grey, smudged overalls. Checked flannels. Scuffed work boots. Even a tattered black baseball hat, stitched with the letters: Mashkin's Plumbing. This, after all, was to be Everett Scarborough's disguise today. That was the plan, anyway: Myers would get them on base as a couple of expert plumbers come to work on a troublesome latrine. It had been Scarborough's idea and Myers had lapped it up immediately as being perfect. And Scarborough had to admit, these grubby clothes certainly were the final touch to a complete overhaul in his looks. Yes, when Marsha Manning had gotten a load of him this morning, she'd laughed—but she too had to admit: it was perfect. Just perfect.

Marsha. The only pleasant thought in his mind. She lingered with him now like a beautiful melody. The memory of last night was a quiet and astonishing joy. Marsha—

"So where's the spook?" Mashkin asked through a mouthful of food.

"Shhh! Walter," said Scarborough in a low voice, "this is a public area!"

"Oh, sorry. Still and all, it's pretty empty in here. Nobody cares much. Besides, how many people know that 'spook' means 'CIA agent'?"

"Walter!"

"Whoops. Sorry." But Mashkin shrugged in a blasé way, as though to prove that this was just another ho-hum day in the life of Walter Mashkin, Top UFO Investigator and Adventurer Extraordinaire. Actually, Scarborough could see now that his loudness and exaggerated movements probably meant that the man was nervous as hell, but trying to act cool and macho about the whole thing and not succeeding very well. "It's just that the guy said he'd be here at 9:15 A.M."

"Or by 9:30. Don't worry. He'll be here."

Or Scarborough hoped so, anyway. His stomach did a nasty little flip as he watched a dollop of maple syrup splash down onto Mashkin's plastic tray. He picked up the coffee and drank some, turning away, turning his thoughts back to Marsha to keep his own form of nervous jitters in check.

Last night had been the single best time he'd ever spent with a woman, sexually anyway. Scarborough didn't know—maybe it had just been his emotional need coupled with the tension that they'd both felt that made it so explosive. But it was more than that, surely—they had blended together so well. It had been so sweet, so fulfilling, so exciting.

Marsha Manning, of course, had her own agenda this morning. Support, she called it. Flanking support. A trip on her own to Kirtland, to scope out the territory, and to create a safe harbor if they needed it. They'd decided it should be the Officers' Club. "That way," Mashkin had said, "if we get holed up for a time, at least we can wet our whistles!"

Ed Myers walked through the doors of the McDonald's at exactly 9:29.

He was wearing a light blue poplin jacket, a red tie loosely knotted around a white shirt, and well-shined loafers. He looked good, relaxed, and tan, with only a trace of worry-wrinkles detectable when he took his sunglasses off to look around the room.

"Ed!" called Scarborough, waving. "Ed, over here!"

Myers smiled, essayed a quick salute, and strode over to them. Scarborough stood up, and when Myers offered his hand, he couldn't help giving him a hug.

"My goodness. Is that a plumber's helper in your pocket or are you just happy to see me!" said Myers, ever the blithe joker.

"It's *great* to see you, Ed. Are you kidding me! Oh, and this is Walter Mashkin."

The two shook hands.

"Are you two ready for a little plumbing expedition?"

"If you think it will be fruitful, Ed."

Myers looked around with a professional assaying glance. "I think maybe we'd better discuss that outside, hmm?"

"Hey, I'm not finished my breakfast!"

"I'll buy you the lunch of your life, Walt. Now come on."

They left the shaded, fat-and-wax scent for the bright New Mexican parking lot. Scarborough put on his hat and his own sunglasses. Mashkin just squinted obstinately.

"That's your pickup over there?" said Myers.

"One of a fleet."

"Fine. That's my Lincoln there. You know you guys, I had a hell of a time manufacturing a classified sewage problem, but I have to admit, this is a good cover."

"You'll be taking us out to the compound where Diane is, then."

"That's right. And you'll have clearance badges. The works. Like I promised. Only then, I'm afraid you're going to be on your own, because if I get caught helping you, Ev, you can well imagine what's going to happen to me."

Scarborough clapped a hand on his friend's shoulder. "You don't know how much this means to me, Ed. This is the end of a long, painful journey. I just wish we had time to talk more."

"You work this out, Ev, and we'll have lots of time to talk. And I'll be buying the beers."

"Oh, I'll work it out. Like I say, I'll have no problem proving my case. But I want to get Diane out of there first. I don't trust your bosses, I'm afraid."

Myers grunted. "Strange business. Well, now, here's the plan, then."

They followed Myers's Lincoln in Mashkin's pickup to the main gates of Kirtland Air Force Base. There were other gates, closer to the far end of the large base to which they were headed, but those put more scrutiny on clearance badges and such. No, Myers had assured them. The main gate would be the best, and then they could use the base roads to get back to the CIA compound where Diane was being held.

Ed Myers had given them clearance badges and credentials. The previous day, Mashkin had worked up employment papers for Scarborough, who was operating under the name Ted Phillips. The few miles to the wide gates of Kirtland were a fast and easy trip. The Lincoln stopped at one of the stations between the lanes, in front of a striped guard-device. Myers got out of the car, showed his credentials to a uniformed man, and then gestured back to the pickup truck. The uniformed man—a corporal, Scarborough could see by his insignia—nodded and motioned for Myers to drive his car through. He went back into the station and hit the switch for the lever to open for him. Myers drove through, then idled, waiting for the pickup to be passed.

"This is it, good buddy," said Mashkin, his Tennessee twang much more obvious now with his nervousness. "Here we go. Yeah, Lord, though I enter the Valley of the Central Intelligence Agency, I shalt fear no evil."

Mashkin drove up and stopped by the corporal.

The Air Force man was one of those young men whose shorn hair, five years in the service, and starched uniform made older looking. He looked hard as nails at first, and for a terrifying moment, Scarborough thought that surely such a professional man would recognize him instantly.

However, the corporal didn't even look at him.

"Passes and prepared documents please," he said, hard as nails.

"Shore thang," said Mashkin, handing the stuff over.

He had a square face, and a sprinkling of acne scars, this young man, and his eyes flowed over the documents with careful and high-trained scrutiny. He lingered on them for an excruciatingly long time. Scarborough felt his heart beginning to crowd into his throat.

Finally, the corporal handed the papers back. "Looks in order. Must be a serious problem to bring in outsiders."

Mashkin grinned. "Yep. That's what I told Mr. Suit and Tie up there. That's why we brought in the best goddamn pipe man in New Mexico." He jerked a thumb toward Scarborough. "If this guy can't get the mail movin', nobody can!"

"Well, that's good to know. I hate to think about officers' latrines being backed up!"

As the corporal motioned them onward, Scarborough heaved a sigh of relief.

"I tell you, friend," said Mashkin. "Ain't officers' shit I was worried about back there."

"Tell me about it!" Scarborough shivered, eyes held steady on the spread of dull-looking buildings ahead. In the distance came the scream of a jet approaching the air strips. "I thought *we* were the ones who were going to get flushed!"

Mashkin exploded into donkey-honks of laughter. He slapped Scarborough on the shoulder. "That's the spirit! I didn't figure you could hang around Walter K. Mashkin without gettin' some Tennessee humor mixed up in your blood."

"What, *outhouse* humor?"

"Scarborough! Is there any other kind?"

Scarborough managed a faint chuckle before the leaden seriousness clamped down on him again.

The Lincoln made a left-hand turn onto a main thoroughfare just before the first clump of official-looking buildings, and speeded up. Mashkin had to stomp on the accelerator to keep up. "Hell! What does that asshole think I'm packing beneath my hood, a V-8 fuel-injected? Kee—rist!" His hands felt their way to the glove box, opened it, and pulled out a bar of Red Man Chewing Tobacco, of which he proceeded to take a health chaw. "How many miles away he say this place is?"

"About twenty, I think."

"You think the government's got enough land out here in the West? Tarnation! It's like they're an occupying force or somethin'!"

"The whole government."

"Huh?"

"I said, that that's the way the government's getting to be—the more I get know about what's been going on, Walter, the more I respect the intentions of our Founding Fathers."

"Shee-it. Tom Jefferson and Ben Franklin would take one look at Washington these days and have immediate heart attacks. You really can't compare the times, fella."

"Oh yes you can. And we have the Constitution and the Bill of Rights to uphold those principles. Just remember, *they're* the law, Walter. The military and the cops? Just hired whores with unconvicted criminals as their bosses."

"Yikes. You getting cynical on me?"

"Maybe."

"Scarb, good buddy. You can't be that way. You got too

much going for you!'' Mashkin let go a healthy spit out the window.

"I let them pull the wool over my eyes for over twenty years. For two decades I was a pawn—a victim of my own enormous ego. Well, I've changed. And if I get out of this thing alive, the world is going to know about what I've been through!''

"Oh, yeah, right—that's A-OK, fella. You just do that, and make sure that Walter K. Mashkin gets himself a footnote or two. But that's not what I'm talkin' about! I'm talkin' about your attitude. I mean, basically, I could have told you about this *years* ago. Your attitude really sucks, pardon my French. And you may have changed your mind about some things, but it sounds to me as though your attitude *still* sucks!''

"My God, Mashkin. Would you care to give me some kind of small break? My whole world has been torn apart, my daughter is the prisoner of unregenerate fanatics—''

"Oh sure, and I'm supposed to feel sorry for you, huh? Well hey, I had a kid once, too. A little boy. Yeah, and I had a wife, too. An automobile accident took them away, and my world kinda deflated for a while, too.''

"I'm sorry. I didn't realize . . .''

"That's okay. I never talk about it. What's the sense. What I wanna say to you now, buddy, is that I kept the faith. I didn't give into cynicism, hard feelings. You see, that's one of the reasons I'm such a UFO nut. I guess it's my way of looking up at the stars and touching 'em. You know—find out the mysteries of the universe, of God, or whatever God is supposed to be. And hell, maybe myself in the bargain. So yeah, maybe you're right in your books. Maybe this whole credulity thing has gotten out of hand . . . turning into a religion . . . whatever . . .

"But Scarb . . . Don't you think that maybe there's somethin' in human beings that yearns for somethin' more that *makes* us that way. Don't you ever just sit by a nice blue lake and look up at the big night sky and just kind of shiver and say, 'Wow! Life is really somethin'. And I sure as hell would like to be closer to what it really is.'

"Well now, I guess maybe that everybody's got their ways of finding that out . . . Their . . . meditations. Yeah. That's the word I'm gropin' for! Their meditations! Thinking and reading and finding out about UFOs and wondering about what kinda beings could be behind the wheels there . . . Hey, maybe that's my form of meditation, my way of saying, Shucks, universe, or shucks, God, or shucks, whatever . . . I'm crazy about you, and I just want to know *all* about you!''

"A commendable attitude, I suppose,'' said Scarborough, a trace of the old tartness back in his voice. "But if you want religion, why don't you just join some church?''

"I ain't no churchgoer. I had a bellyful of Baptist bullshit up to age sixteen, and I just can't take it no more. That don't mean I don't believe in God. Shit, I even pray to him once in a while. Kinda comfortable, just conversing with the Almighty. No, I'm a loner, Scarborough. I guess maybe you can understand that. I need to do things my way—I don't like getting my rough edges scraped off by no cookie-cutter of dogma and convention. But that don't mean I'm no cynic, Scarb. An' I'm just sayin' that maybe you'd better watch yerself. 'Cos cynics—man, the real ones—they ain't good, they ain't bad, they ain't nothin'. Their hearts may be pumpin', but they're as good as dead inside.''

"Okay, Walt. I'll think about that. I don't feel very dead inside. I'm feeling rather alive. Maybe it's not cynicism, maybe it's just hurt.''

Mashkin winked at him. "You bet. That's the spirit. And those bastards—they did hurt you plenty. But I'll tell you, bud, we're gonna get that little girl of yours outta there today, and then we're gonna find out all about that crashed saucer in Roswell. And maybe, just maybe, I'm gonna be able to meet up with an ET!'' Pure enthusiasm beamed from the broad grin that ignited upon Mashkin's face.

"Let's just concentrate upon the first part, okay?''

The countryside sparsed down from buildings to long stretches of arid land. Off in the distance, Scarborough could see Air Force fighters and carriers swooping in for landings, or taking off in muted blasts of sound; but the road they travelled skirted the runways by a considerable margin.

The drive was another fifteen miles. Scarborough prepared himself inside for what awaited him. Below the seat, of course, they had some of Mashkin's weapon collection: a shotgun and several automatics. He hadn't told Myers about these; but

Mashkin had insisted upon them, and Scarborough had to agree, having them around would not be a bad idea. Anyway, very seldom were there serious searches made of vehicles at the main gates of Air Force bases in New Mexico, especially not if a guy with the clout of Edward Myers got them in. Scarborough just hoped that they didn't have to use the weapons. The theory was that they'd work on the plumbing, then sneak off, find Diane, maybe check out what was going on at the compound. Then, they would hide Diane under a cloth in the back and vamoose on back to the Officers' Club, hopefully without pursuit. Once there, Walter Mashkin could just drop them off and get off base under his own power and papers. Marsha would take care of the rest then, and they'd be in safe waters.

Scarborough was leaving it up to Myers to extricate himself from any problems his activities might create. Myers had served in the CIA for close to a quarter-century. If anyone could get themselves off the devil's pitchfork, it was Ed Myers. Probably had his course of action all planned, yes indeed. An amazing guy, Ed. Scarborough was just happy that he had a good an insider in the system—and as faithful a friend.

They passed through even scrubbier territory, filled with sagebrush and the occasional tumbleweed. The wind that fluted through the window was full of the by-now-familiar New Mexico tang. The taste of that awful McDonald's coffee still clung to Scarborough's mouth stubbornly.

"You don't have any chewing gum, do you, Walt?"

"Got some Juicy Fruit in the glove box there. Help yerself!"

Scarborough clicked the box open, and was almost deluged by the cram of stuff packed inside. How Mashkin had gotten his tobacco out without dumping it all on the floor he'd no idea; right now, he had all he could do to prevent that event. There were maps, old car registration cards, scissors, tools, receipts, and what-all-else here, and it took a little rooting to find the pack of gum that Mashkin had promised. He *did* find it though, and it proved to be a fresh megapack. Scarborough pushed the rest of the stuff back into the dashboard cabinet and unwrapped two sticks of gum. The sweet taste made him feel immediately better; they even seemed to settle his sour stomach somewhat.

"Yep. You can't go wrong with Juicy Fruit," said Mashkin.

"That's my favorite. Backslid into Doublemint for a while, but that Juicy Fruit is my once and future gum!"

"When you're not chewing tobacco."

"Shit. Just when I'm nervous. This stuff tears hell out of your mouth, you do it too much!"

"A notice from the surgeon general!"

"You makin' fun of me, boy!"

"Impossible, Walt. You're already a barrel of laughs."

"Well, I'd advise you keep a tight hold of your trousers, 'cause the yuks are just gonna keep on a comin'." He pointed up ahead and Scarborough looked.

It was the Complex.

Just as Myers had said, it was in the middle of nowhere, looking somewhat like an afterthought tacked onto the military base. Indeed it was like a microcosm of the base, the same tired architecture, the same bad paint jobs, the same humps of hangars and blocks of buildings.

Only this little CIA enclave had its own private fence, its own private gate, its own private guards.

"No problem," said Mashkin, pulling his papers back out from the sun visor. "We made it through one; we'll make it through this."

"Yes," said Scarborough. "Of course we will." Had Myers told him about this gate? He must have. He just couldn't recall it offhand.

It really was a perfect place for a semisecret CIA installation. A middle-of-nowhere spot on a high-tech-defended Air Force Base in the middle of nowhere. And even though they were miles away from the main base, these buildings blended in with the others of Kirtland. No way would anyone want to investigate them. It would be tough to even find out about them, and even if they did, how could one get through a double set of gates with the Air Force protection? Yes, a good place to set up shop if you're going to do hideous things to victims of kidnapping, perform awful experiments; a good place for doctors to break the Hippocratic oath on innocent victims in the name of national security.

Myers's car stopped at the gate. A uniformed guard stepped out to check his identification. Myers waved for Mashkin to stop his pickup truck and Mashkin did. Myers walked back to them, still smiling but looking a little tense, nonetheless.

"I was afraid this might happen, but it's really no problem," he said as he leaned casually into the driver's window.

"What's wrong?"

"Oh, nothing. Nothing. Routine check, that's all. They want you to get out and come up to the booth with the papers I gave you. I'll handle the whole situation. No problem at all."

Mashkin shrugged and got out. "Hell, why not? This *is* the American military we're dealing with."

Scarborough was a little less sanguine. Myers hadn't mentioned the possibility of this kind of activity at a checkpoint. Still, Scarborough had worked for the Air Force quite a few years and he knew their security could be a little quirky. The kicker, of course, was that this was really a branch of the CIA here, and *their* security could be tough. Still, Myers wasn't a stupid man; he would have prepared for just such an eventuality.

They walked to the guard station, a booth with windows, much like the ones at the main gate, except that it was longer and deeper with some sort of room with no windows behind it.

"Good morning, gentlemen," said the corporal, a near clone of the one back at the main gate, right down to the shiny buttons of his blue uniform and the razor burns on his neck. "Could I see your documents? We don't usually call on outside plumbers around here."

"Well, sure enough!" said Mashkin, grinning and handing forward his papers.

Scarborough did the same.

Just as their hands were extending outwards, a group of three men in coats and ties and sunglasses stepped out. Two held automatic weapons. Scarborough did not recognize them, but he did recognize the third.

It was Brian Richards.

"Hello, Scarborough," said Richards. "Would you and your friend please put your hands on top of your heads and come along with us?"

Scarborough's heart lurched. He felt as though he was about to vomit, but he rapidly regained control. He swivelled toward Myers, working his mouth, unable to get anything out but gasps.

Myers was looking away, his head bowed. "I'm sorry, Ev. I truly am!"

"What the shit!" said Mashkin. "I'm an American citizen. You can't do this to me!" Nonetheless, the man put his hands on his head.

"We're officers of the Executive Branch of the United States Government." Richards flashed a badge at Mashkin, his face deadpan as Jack Webb's in "Dragnet." "You are hereby under arrest for aiding and abetting a wanted criminal." He turned to Myers. "Who is this bozo, anyway?"

"The plumber. I told you. Walter Mashkin, the plumber."

"Oh yes. The plumber UFO-freak, that's right. I have a file on you, Mashkin. I just didn't peruse it fully." He gestured toward the buildings. "All right, Scarborough. I want you to start walking toward that nearest building, where you'll be detained until we can interrogate you."

"Hey, don't we get Mirandized? Don't we get to call our lawyers? What the fuck is going on?"

"Mr. Mashkin," said Richards harshly. "You've just stepped into a very deep pile of crap. And if you ever want to get out of it, I strongly suggest you cooperate."

The shock of the reality of his situation lost its paralyzing aspects in Everett Scarborough, and all the rage and fury that he felt toward this man and what he represented coalesced into an uncontrolled reaction.

Scarborough leaped toward him.

"Wha . . ."

The automatic weapons turned their dark nozzles toward Scarborough.

"No! Don't shoot him!"

Scarborough grabbed Richards by the neck. His momentum and his anger drove them both down onto the tarmac. Squeezing for all he was worth, Scarborough tried to choke the life out of Richards. But before he could get anywhere near that goal, he was struck from behind by a heavy, hard object. A stunned moment full of disorientation later, he found himself spread-eagled on the hard ground, a terrible pain blooming in his neck and the back of his head.

He heard the sound of running footsteps and he looked up in time to see Mashkin running toward the cab of his pickup truck. A brief flicker of hope came to him; if Mashkin could make it back to the Officers' Club, tell Marsha Manning what had happened, there still might be hope . . .

"Stop him!" barked Richards. "For Christ sake, stop that man!"

A pause, and for a moment Scarborough thought that Mashkin would make it. He had the door open and was just climbing in.

And then the guns chattered.

Bullets stitched across the old truck, puncturing the worn Mashkin's Plumbing letters. They tore up chunks of flesh and blood across Walter Mashkin's back. The man arched backwards with a cry of pain, his hands fisted up toward the sky—and then he fell, slamming onto the road surface like a sack of potatoes.

"Walter!" cried Scarborough, getting up to go to him. But the government men grabbed him and held him fast.

Scarborough could see his friend was dead, though, now. His eyes were fixed up toward the blue and the clouds of the sky, as though even in death vigilant for another last glimpse of flying saucers.

"You *bastards!*" Scarborough cried.

Richards grunted, unbuttoned his shirt. "Yes, if necessary. We'll take care of that body later. Right now, I want this man searched for any possible weapons and then placed in handcuffs and in a cell. Understood?"

With the assistance of the guards, Scarborough was hauled away.

"I'm going to *kill* you, Richards," cried Everett Scarborough. "As God is my witness, I'm going to *kill* you!"

"You see, Ed," said Richards. "A very dangerous man. You did the right thing."

Ed Myers said nothing.

CHAPTER THIRTY-TWO

Something was wrong.

Lieutenant Marsha Manning sat in the Officers' Club of Kirtland Air Force Base, with the terrible feeling in the pit of her stomach that things had gone terribly askew in Everett Scarborough's mission at the CIA enclave. This wasn't just intuition, either. It was getting on toward twelve-thirty in the afternoon.

They should have been here at least an hour ago.

She sipped her Coke and nibbled at the chef's salad she had purchased so as not to look out of place; people had started rolling into the place for lunch about eleven-thirty, and she would stand out like a sore thumb with just a cola for company. So she sat now alone at a table, with some reports stretched out before her, seemingly immersed in work, and actually immersed in worry.

She'd gotten to the base early enough, and had checked into the appropriate channels. Ostensibly, she was here as a computer consultant, a duty that she had constructed with the help of a chief programmer friend at Wright-Patterson, and heavy string-pulling with communications bulletin-board buddies here at Kirtland. She was glad now that she'd had that lonely time last year, when she'd taken refuge in long-distance conversations through the medium of her computer screen and keyboard. The network of contacts she'd made had already paid off for her, and now it was paying off in a big way. Without these people's help, she would never have been able to set up this "assignment," let alone get off Wright-Patterson for a while.

The computer consultant wasn't supposed to begin until

tomorrow; Marsha had the day to set things up, prepare. But of course she'd used it not for computer purposes, but to set the stage for the rescue of Dr. Everett Scarborough, his daughter Diane, her fiancé Tim, and of course Walter Mashkin, along for the ride. She had to admit, it was a good plan, and a smart idea of Scarborough's. As rotten and corrupt as these Editors were, and as closely tied with the CIA and the Air Force, they were covert; American law still held sway. Scarborough planned to be unarmed once he gave himself up here at the Officers' Club. They would take him, but he would still have all his rights as an American citizen, and he would be under the protection of a higher authority, which Marsha was determined would protect him. She already had a lawyer ready, and she knew just what MPs to call, to say nothing of the press. As soon as Ev walked through that door, she was going to be on the horn to both, and everything would be just hunky-dory.

However, the key event—that selfsame spectacular entrance of the fugitive doctor himself—was simply not happening, and Marsha Manning's stomach was beginning to act up. She pulled a package of Tums from her purse and chewed a couple of tablets. She looked around speculatively at the faces above the brass, eating and conversing.

The smell of the kitchen's special—some kind of veal cutlet—was strong in the air, and the place was filled with the din of cutlery, dishes, and talk. Much the same kind of scene as might have taken place ten, twenty, or even thirty years ago, with the significant difference that instead of martinis or bourbon and branches arrayed about the officers, now there was mostly soda-water and lime. The military was discouraging alcohol intake, much as everyone else was. After spending some time with a Jake Camden or a Walter Mashkin, Marsha could understand why. You didn't want a hard drinker behind the controls of a jet or an Air Force base; the myth of the manliness of drinking was evaporating quick among the military ranks, and thank God for it.

By one o'clock, the place was almost full, and the waiter wondered if maybe Marsha wouldn't like to sit at the bar; they hadn't expected her to be this long, and they'd reserved her table for another party at one-fifteen.

So Marsha Manning went to the bar. She ordered a ginger

ale, no ice. She sipped it. And waited. She didn't know what she could possibly do, until a certain man in a uniform entered.

"Rick!" she called out, waving her hand. "Rick, over here!"

Rick Hawkins was a slender, dark-haired man of perhaps thirty-five with a large Roman nose and a prominent overbite. He was a captain here, who did some work with computers, but his main chore was base construction supervision. Captain Rick Hawkins was one Marsha's principle computer pen pals, and they had met once before, when he'd visited Wright-Patterson. It was obvious that he was very taken with her; he'd bought her dinner and come just short of propositioning her. With her feminine wiles, however, Marsha had managed to keep him at arm's length without exactly discouraging him, so this morning he'd been very happy indeed to see her, and seemed to be even happier now, that they might have leisure time together.

"Marsha! Taking a loooooong lunch, I see."

"I'm—ah—waiting for someone."

"Me?"

"No, though I must say, I'm very happy to see you, Rick!"

He brightened even further. "Have you reconsidered dinner for tonight, then?"

"Yes, as a matter of fact I have."

"Oh, terrific. I'll cancel my bowling, then. How often does a guy get a visit from a beautiful and brilliant young woman he can actually gab heart-to-heart with, via the miracle of telecommunications?"

"Not very often, Rick. Yes, I'll have dinner with you tonight. But a friend of mine hasn't shown up for an appointment, which means that I'm going to have to go and see him at his office, and that's quite a distance."

"You need a ride?"

"No. I'm going to be awhile and I need to come back." She paused a moment, looking deeply and sincerely into his bluish grey eyes. "I was hoping to maybe borrow that car. What did you say it was—a Toyota Celica?"

He smiled. "Yes, that's right. You remember!" He looked around, massaging his chin as though in thought, but Marsha knew that he was just playing a game, playing "Hard-to-get." "Well, I guess I really don't need it for this afternoon. I can just walk back to the office. Oh hell, yeah. Why not? That way, I'll be able to see you for more than just dinner, huh?''

Marsha smiled and tickled his ear teasingly. ''Rick, you let me borrow that nice car of yours until four, and afterwards, I'm all yours!''

Of course, Captain Rick Hawkins had no way of knowing it, thought Marsha, but the loan of his car could net him a good deal more than just a grateful Marsha Manning for dinner. But then, she had no choice in the matter. She was desperate.

She batted her eyes at him prettily. ''By the way, Rick. Is it at all possible you have a map of the base in your glove compartment!''

CHAPTER THIRTY-THREE

They had him handcuffed and covered by several guns, and they'd shown their willingness to use those weapons by blowing away his friend. So Scarborough had no intention of trying to escape. Besides, he was simply too stunned, too shocked, by the suddenness of it all to take any kind of action. Not now, anyway.

Nonetheless, after they'd marched him into one of the buildings and into a secured room, they hadn't taken off the handcuffs. No, Brian Richards ordered for additional cuffs to be fastened from his legs to the chair in which they seated him.

''You're not being too careful, are you?'' he muttered.

Brian Richards surveyed the situation silently for a moment. Finally satisfied that Scarborough wasn't going anyplace soon, he grunted. ''Pardon me. Did you say something?''

Scarborough said nothing. The surprise was beginning to wear off now, and the grief was beginning to leak from somewhere deep in his soul. The image of Walter Mashkin being shot down like some hunted animal stained his mind in shades of red and blood and death; and he knew it was indelible for as long as he lived. Which didn't seem likely to be very long at this point.

And Diane . . . He'd failed her. Again, he'd failed his daughter when she needed him the most.

All his insides felt clenched tight as a fist.

Richards sighed and sat down in a chair opposite his captive. "This really is for the best, Scarborough."

Scarborough lifted his head and glared at the man. "For you! Why didn't you just gun me down in cold blood, like you did to Walter Mashkin."

"Was that his name. Mashkin . . . Mashkin . . . Oh, yes, of course. The saucer afficionado from Albuquerque. We really should have guessed from the very first . . . Oh well, it was a shame to kill him. But he tried to escape. Something which you were wise enough to refrain from." Richards cleared his throat, stood, and stepped over to Scarborough. He cuffed the man's ears much as a schoolmaster might a wayward student's in the nineteenth century. "Because I swear to you, Dr. Everett Scarborough, if you try to escape before we're done with you, we will *have* to kill you." There was nothing in his slate grey eyes but stone and death.

"What have you done with my daughter, you bastard!" spat Scarborough, swallowing back the pain and standing strong against this intimidation.

"Daughter. Who, you mean Diane? You really think we've got her, don't you?"

"Who else would have her?"

"So tell me. Suppose it all comes out that you get your beloved Diane back, Scarborough. Will you cooperate with us?"

"I'd rather burn in hell!"

Richards laughed. "That can be arranged, believe me." He stretched, cracking his knuckles in a way more banal than ominous. Keeping his face in its hard composure, he looked away, as though considering the situation deeply. Scarborough took the time to take note of his surroundings.

He was reminded of a doctor's examination room. The room was basically an off-white color, and there were cabinets that might well contain a doctor's instruments, a doctor's vials of drugs. There was no examination table however, and no telltale stainless-steel contraptions or stirrups in evidence. *Neutral* was the word that came to mind—but fast on its heels came another word, more intuitive: *deadly*.

"I think you can leave us now, gentlemen. I don't think he's going anywhere."

"But sir—" said one of the suits. "You said he's quite dangerous and shouldn't be left alone."

"So I did. Very well, please post yourselves outside the door. but first go and notify Dr. Cunningham that she's needed."

"Yes sir."

The men turned and left, leaving them alone.

"Now then, I quite assure you, we don't have your daughter, Scarborough. Her disappearance is quite the mystery to us."

"I don't believe you. And you're about to tell me that you didn't kidnap her fiancé?"

"Mr. Timothy Reilly? Why, of course we did."

"What have you done with him?"

"Actually, he's in very good shape and just going through the last of his finishing touches before we let him venture out into the world. You see, Mr. Timothy Reilly up and decided to fly to Europe. To get away from it all, you understand. His parents have been receiving regular postcards from Munich, Florence, Venice, the French Riviera—all the usual touchstones of youthful peregrinations. And old Reilly is really very pleased about the whole thing. Seems that Tim has dumped your daughter and has straightened out quite a bit. Wants to go into the old man's business, once he's sown the last of his wild oats."

"That doesn't sound like Tim! What have you done with him?"

"Oh, suddenly you *like* Tim Reilly's personality? He was sowing quite a few of those wild oats in her, as I recall, and you were not particularly pleased."

"Diane's her own person. You didn't answer my question."

"I don't have to answer anything, Scarborough. I'm the one with the questions." He put his face right up against his captive's and fairly snarled directly into it. "Why did you betray us, Everett Scarborough! Why did you betray your government!"

"I've betrayed nobody. I am loyal to my country. Illegal portions of that government have used me for their own ends. I mean to *expose* that!"

"Oh yes, and you're doing such a good job, aren't you? No, Scarborough. You betrayed us. Surely, deep down, you realized that your fame and fortune were because you played the sycophant for a greater power than you. In a worthy cause, a *worthwhile* cause—national security!"

"Am I going to find out *why*?"

"I'm sure you have your suspicions. But that's neither here nor there. What we need to know, *now*, is the extent of the damage you've caused to our project."

"I don't know. Lots, I hope."

"Hmm. Stubborn. I thought you might be. No matter. You will be dealt with."

"What? With torture? Drugs?"

"Please. That's not my department. But I will say that you will make it a lot easier on yourself if you cooperate. Tell us the entirety of your activities."

"You know I didn't kill Mac MacKenzie."

"Oh, of course. Our operative had to kill the captain. Rest assured as soon as this matter is cleared up and we feel you are again a trustworthy member of our group, you will no longer be a hunted criminal—not even for the death of that operative at Hoover Dam. You see, we can be quite generous."

"And if I won't cooperate?"

"Oh, you will, Scarborough, you will. But if we think you're still a wild card . . . Well then, I suppose your suspicions are entirely correct. We'll just have to kill you."

"For reasons of national security."

"That's quite correct."

"'My country 'tis of thee, sweet land of liberty.'"

Deep frown. "I do the sarcasm around here."

A cold anger swam in Scarborough's veins, a deep and powerful strength that he did not know he possessed. After all these moments of free-floating anxiety, all the fear and terror—now he was at rock bottom, caught and helpless in the web of his enemy, he now felt a strength that he did not know he possessed. A strength not of desperation, nor of nihilistic resignation—but of character.

"Do what you want to. You'll not get anything out of me while I'm still me."

"Lovely. I'm sure Dr. Cunningham will enjoy herself, then. Oh, by the way, we've got your friend Jake Camden. He cracked in very little time at all. Looks like we're going to have another very cooperative journalist to work with!"

"Fuck you!" shouted Scarborough.

Richards coldly slapped him across the face.

Scarborough could taste blood. He let his head hang for a moment, then sucked up the saliva and blood and phlegm, swung his head up, and spat the mess upon Richards's clean shirt and jacket.

"Jesus Christ!" said Richards, and he hit Scarborough again, harder. "I wish we did have your cunt of a daughter, you asshole. Then we'd control two whores from the same family!"

The world seemed to go black. Scarborough lunged with all his might toward his adversary. Richards stepped back, allowing the chair to collapse upon the floor, with Scarborough in it, painfully sprawling as far as his confinement would allow. Frustrated, Scarborough lay there, desperately working against the impossible, to free himself and grab his tormentor by the throat. Peripherally, he heard someone enter. The female voice that followed somehow stilled him.

"Well, I see you boys have been quite busy in my absence," the woman said with total contempt.

"I'm afraid our Dr. Scarborough has a bit of a chip on his shoulder," said Richards, taking out a handkerchief and cleaning himself off.

"Obviously. And your usually calm temper has flared somewhat. Well, I'm not going to be able to do anything with him lying there. Maybe you'd better call your goons in and set him up for me, hmmmm?"

Richards summoned his men, and Scarborough felt his chair being righted. The faces of his captors swam into view dizzily: Richards, still mopping himself; and a blonde woman in a lab coat and glasses, hands folded under her breasts in a stern and almost schoolteacherly fashion.

He recognized her. He'd met her before, at a few social functions involving Colonel Dolan, and he'd seen her at conferences he'd attended for the Bureau of Mental Health. The Ice Queen, he'd heard her colleagues call her behind her back. Dr. Julia Cunningham. He was right then: he'd been right in his suspicion, his guess, after what the woman in Baltimore had told him. This was the principal doctor working on the fake alien-abductee program. This was the woman who held in her head and her little black bag of medical tricks secrets of mind-control unknown as yet to the general medical community.

"Hello again, Doctor. I don't suppose you have any medicinal alcohol about? I could use a stiff drink," he said, his emotions coming under control.

Dr. Cunningham ignored him. "I trust if he's going to have strong tendencies toward violence that I have use of your men for the time being."

"Yes. No problem there." Richards examined his watch. "I'm afraid you're going to have to do without me. I have to catch a plane very soon. Some essential details I have to deal with back in Washington. But I'm very happy with the turn of events here. Very pleased. You'll report to me on your progress with—with our guest?"

"I trust you'll change your clothes on the plane. You look terrible, Richards. And as to Scarborough, here." She turned to him, raising an eyebrow. "I shall extend full professional courtesy to Dr. Scarborough. And I think, that, given a day or two, he shall indeed be a new man."

A hint of a glacial smile touched her mouth and Scarborough could feel frost on his spine.

"Excellent. Scarborough, I commend you to our good doctor's healing hands. And I hope that our conversation can be more cordial the next time we meet."

"The next time we meet, I'm going to kill you, Richards. I swear it."

"You see, my dear," said Richards with an uneasy chuckle. "Somewhere along the line he's picked up quite a bit of hostility. Deal with it, please." With one more hard glance at Scarborough, Richards stormed from the room.

"Dr. Scarborough," said Cunningham. "My colleague tends to banter. I do not. I perceive resistance on your part. I warn you in advance, resistance can be broken, albeit painfully. I suggest that you relax and comply as best you can with the situation."

"Thanks for the warning, but I don't think that's going to be possible."

"No? How unfortunate. It's seldom that I'm allowed to deal

with an intellect as strong as yours, a personality as complex. When the patient is not compliant, stronger measures, stronger doses, as it were, are sometimes necessary. This tends to occasionally have unfortunate consequences."

"I thought you didn't banter, Doctor."

"In short, Scarborough," she said through gritted teeth, her eyes flashing with ice. "If you want a shred of yourself left after I'm through with you, you'll cooperate."

Instinctively, Scarborough realized he was going to have to change tack with this woman. Richards was clearly an individual who, beneath the hard surface, respected hardness in others. This woman was clearly all business. A different sort of game was clearly in order.

"I don't understand. How can I cooperate? You've got me tied down. You can do whatever you want with me. What difference does it make, Doctor?"

"Are you volunteering cooperation?" A hint of suspicion edged her voice.

"I'd like to know what it is, first."

She snorted. "An attitude, of course. Surrender, if you will. Resignation to your fate."

" 'To lay me down with a will.' "

"Yes. Exactly. Naturally, part of the process is the extraction of information."

"Right. Are your methods similar to those used on William Buckley in Beirut?"

A slight smile. "Barbaric methods, those, but ultimately the information was extracted. No, Scarborough. We are light-years ahead, now. There is no actual physical torture involved. Although I am told that at times physical torture is preferable. To the patient."

"So much for the Hippocratic oath, eh?"

"I am a doctor of science. I serve my country. I have no apologies. But you've changed the subject. We merely want to obtain certain items of information. They can be easily verified, so telling the truth is suggested." Puzzlement creased her brow. "But Scarborough, you refused Richards. Why the change of heart?"

"Maybe I've had time to think."

"You'll talk then?"

"Just tell me what you want to know, and I'll do my best."

"Hmmm." She shrugged. "Very well, let me get my tape recorder." She went to a side table and pulled open a drawer, extracting a small portable recorder. She inserted at tape, then turned it on.

"Does this mean no drugs?"

"Would that disappoint you?"

"Did anyone ever tell you you might be the prototype of Lilith Sternan on 'Cheers'?"

" 'Cheers'?" What's that?"

"Never mind."

Yes, he'd answer questions. Because he knew that once they started inserting the needles, once they emptied their terrible chemicals in his bloodstream, that would complicate things tremendously. He had to stall. He realized there was one hope, and that was Marsha. When he and Mashkin were late, she might be able to come out and investigate.

That was the hope that he had to cling to.

"Go ahead. I'll do my best."

"Excellent. Dr. Scarborough, I have been examining your records and your tests for some time." She was silent for a moment, as though formulating the precisely correct phrasing. "A good many records exist, of course, concerning you in light of your role in Project Blue Book."

"You mean, my role as patsy."

"Whatever. Of course, you can well imagine the file that we possess on you."

"Extensive, I dare say."

"Yes. But to begin the questioning today, Dr. Scarborough, I shall come directly to an area that causes me great consternation. To wit: there is an element of your personality that has always puzzled us. But allow me to frame my question in this manner:

"Everett Scarborough, do you have any reason to believe that at some early point in your life, you may have been undergone an alien-abduction experience?"

CHAPTER THIRTY-FOUR

Ed Myers flushed the toilet and then just knelt there, before the commode, sweating, letting the sickness, the pain, the guilt, pass over him like a wave. The vomiting had helped somewhat, but shreds of the nausea still clung around the back of his abdomen as though hooked upon his spine.

I'm sorry, Scarborough, Myers thought. *Jesus, I'm so* sorry.

But he *had* to do it. They hadn't left him any choice in the matter. They'd threatened his *family* for God's sake, and he knew with every fiber of his being that when Brian Richards, the unholy bastard, said something, he meant it.

Eventually, Myers got up and went to the sink, where he splashed cold water in his face. He dripped for a moment, and then grabbed a towel and wiped his face. Reluctantly, he lifted his reddened face and stared at himself in the mirror.

His eyes looked as though they were retreating to the back of the skull, and what had once been modest wrinkles now seemed more like fissures. He didn't look as awful as he felt, though—which was like the *Portrait of Dorian Gray*. For the first time, he was confronting what this business had turned him into.

Sure, on the surface it had all seemed just fine, thank you. His family was the ideal Reagan nuclear unit, the ideal that conservatives flashed across millions of TV screens, pushing for Ronnie's election and re-election. His job was the ultimate in "ask-not-what-your-country-can-do-for-you" professions. He was an Ivy League college graduate; he'd served a two-year hitch in the marines; my stars, he even fucking flossed his teeth, stopped smoking after college, stopped drinking after his first child was born, kept a good attitude, a healthy American optimism. He'd refused to kill people; he's made sure that his job as CIA operative had never put him in that position. He'd never ordered anyone killed. His duties had stretched from Chile and Central America to Beirut and the Far East, to say nothing of the delights and extravagances of Europe. And yet, even though he was separated from her for long periods of time, he'd never cheated on his wife, Carol. Despite his absences, he tried to be the best father he could—fair, reasonable, and loving. He'd seen the things that power could do to a man, and yet somehow he'd resisted them. He'd thought of himself as a good, just, dutiful citizen—and yet, now, as he looked at himself in the mirror, he knew that it was all rationalization; that throughout his twenty-plus years of service, the rot had slowly crept into him, and now it had broken out in full force.

He had betrayed his friend, he was in the Judas Iscariot league now. He was just a fucking cipher, a cog in the machine, and when that cog had jammed, they'd taken a power wrench to him and he'd worked just fine. Like a drone. Like a fucking drone. So much for ideals, so much for individuality. Behind the fabric, the vaunted democratic republic of this country, parading the noble words of the Constitution about in disguise, there was the skeleton of simple fascism. People like Richards were the skull beneath the skin. And all this time, he'd just been a dumb minion who'd bought the party-line.

Ed Myers let go a sigh in a racking heave. He shook himself and took another towel and got the last of the moisture off his face.

What were they going to do to Scarborough?

The question gonged in his head like a bell. Suddenly, he just had to know that answer. Suddenly, he realized that everything they'd told him, all this gobbledegook about Editors and Publishers and aliens and what-have-you wasn't necessarily the truth.

What were they going to do to him?

Myers put on his jacket. He combed his hair. He straightened his security badge. He donned once more the sheen of professionalism, of CIA control, of glib competence. Last night, unable to sleep, he'd taken a walk around these halls. Last night, he thought he had heard screaming from one of the rooms. He well knew Dr. Cunningham's role in this project, but he did not know that it involved screaming. Now that they had Scarborough—was that his destiny as well? Torture? It was hard to imagine, but he knew all too well the sorry history of the CIA in such matters. When it came down to brass tacks, when matters of "national security" came up, the Americans could be just as down and dirty as the Russians, the Chinese, or, for that matter, the Third World political witch-doctors dragging information from prisoners with cattle-prods and dentist tongs.

And now they had Scarborough. The man he'd betrayed. The man who had helped save his son from drugs.

He had to find out what they were going to do to him. Suddenly, nothing else mattered.

Myers washed his mouth out, stuck in a Tic Tac, and was ready to go.

And he knew just where to go.

The halls had been practically empty. No guards stood at the doorway, which was locked.

However, it was not a difficult lock to deal with. Ed Myers drew out his ring of keys, upon which he had a standard issue lock-pick device. It was one of Myers's specialities, and he had the tumblers tumbling and the knob twisting in just a few short maneuvers of his hand.

The door opened into darkness. The light from the hallway cast only vague shadows from the still forms in the large area. There was a funny mixture of smells in the room; Myers noticed it immediately. There was disinfectant and medicinal scents, but there was also the taste of blood and vomit and other body wastes. It was a terrifying combination, and it squatted in this room like the ghost of a man who had died hideously.

Myers fumbled for the light switch. He found it, but closed the door behind him before he turned it on. Unexpectedly, fluorescent lights did not ignite. Instead, softer side-lighting fluttered on—subdued, almost atmospheric.

Enough to see what Myers had half-expected, but had hoped not to see.

The man was strapped onto a table with leather belts. It was like something out of a thirties horror movie with Bela Lugosi and Boris Karloff, only infinitely more horrible for its pathetic banality. The man was unconscious, but still alive, breathing shallowly. Bloodstains ran down from his nose, vomit stains were on his neck and chin. His clothes were soaked with sweat. This was the awfulness that Myers had smelled upon first opening the door.

Ed Myers did not recognize the man, but he suspected who it was. He stepped to his side and was about to nudge the prone body in the arm, when he saw the worst part.

He saw the needle tracks in the man's forearms.

Jesus. Jesus Christ, they must have thought this poor guy was a pincushion!

And this was what they were going to do to Everett Scarborough; no doubt about it.

The man's skin on his neck was cold and clammy, but the pulse was strong. Myers shook his shoulder slightly. No response. Shook it a little harder. The man's eyelids fluttered. The mouth worked.

"Don't say anything," said Myers. "Don't move. Stay very still for a moment."

The eyes opened, and flashed immediately with absolute panic. Myers could see the scream starting from deep down in the man and coming up fast. He put his hand over the mouth and muffled it.

"Listen. I'm a friend. Are you Camden? Jake Camden, Everett Scarborough's friend."

The man grew quiet. His eyes rolled slightly and eerily as though they'd become loose in their moorings. "Camden," he muttered. "Jake . . . Jaaakkkkkkkkeeeeeeee . . ."

"Yes. Is that you?"

"Make her stop it . . . stop it . . . I swear, that's all I know. I swear I'll do anything. ANYTHING!"

"You're going to be okay. Just settle down. All right?"

My God, what kind of devil's brew did they inject this poor guy with? His hands were shaking as though touched with palsy, and his face was starting to twitch. His eyes started now as though he were watching the ceiling erupt with demons and hobgoblins.

"Don't let them get me! Don't let them GET ME, please!"

"Steady there, pal. They're not going to get you. Now tell me . . . You're Jake, right?"

"Yes . . . Jake . . . Jake . . . That's me. Jake . . ." Pause. A deep sigh. "Jesus, I could use a drink. They don't have booze on flying saucers, though. No booze . . . no beer . . . Drying out . . . Need a drink. Jesus!"

"I'll get you a drink. Just settle down, okay."

The very thought of a drink seemed to calm the man somewhat.

"Okay, Jake. If you promise not to do anything weird, I'll let you out of these straps."

"Yeah, yeah . . . straps . . . God, they hurt. They hurt bad."

They *were* ridiculously tight. Christ, he had the reading on this Cunningham lady all right. She was a goddamned *sadist*. Leave it up to Brian Richards to recruit a sadist to head up a program. It took awhile, but he got the straps unbuckled. He helped Camden off the table.

Camden collapsed onto the floor.

"Oh God, they cut off my *legs*!" he moaned from the ground.

"No, they didn't cut off your legs, man. The circulation's been cut off."

"Well, *something's* sure been cut, that's all I know."

Myers rubbed Camden's legs and arms methodically, helping to restore the circulation. Still the man couldn't stand. All Jake Camden could do was hobble to a chair, where he crashed down, breathing shakily.

"Whew! What did they *do* to you?" said Myers.

"Drugs . . . I don't know what else . . . Oh God . . . I almost lost all identity, all hope . . . Who are you, anyway?"

"Myers. Edward Myers."

It took a moment to register, but suddenly recognition entered Jake Camden's drug-colored eye like a faint glimmer of hope in total gloom. "You're Scarborough's buddy. His spook friend!"

"That's right."

"Where's Scarborough?"

"Do you think you can walk?"

"I don't know, I'll try." Camden wobbled to his feet, lingered there for only a few unsteady seconds, and then collapsed back into the chair. "Yep. I can walk, but give me a few minutes, huh?"

"Stay here," said Myers, starting for the door.

"Yeah. Like I'm going anywhere quick in my shape," Camden's voice was still vague and thick with the drugs. "But you didn't answer my question. Where's Scarborough?"

When Ed Myers turned around, he was holding an .38 automatic gripped tightly in his hand.

You could push a man only so far past what he believes in, Myers thought. And then either he snaps—or he snaps back.

Edward Myers had not snapped.

"I'm going for him right now," he said, and ran out into the hallway.

CHAPTER THIRTY-FIVE

"Kidnapped by aliens?" said Scarborough. He tried to keep up a bluff manner, but his mouth went dry and he stuttered the rest. "Th-that's *your* scam, isn't it, Doctor?"

Dr. Cunningham was silent a moment, studying her captive carefully. "There's no sense in denying that our project has encouraged the development and nature of the phenomenon. But it *is* a phenomenon, Scarborough. Make no mistake about that. It does happen."

"What? You're not trying to tell me there really *are* aliens walking around on earth doing irrational things?"

"That aspect is inconsequential." She tapped his forehead and her finger felt cold on his brow. "I'm talking about what goes on in the mysterious depths of the human mind, Scarborough. I'm talking about pyschology. Carl Jung was fascinated with the flying saucer phenomenon. I suppose you know that."

"You clearly haven't read my books."

"Oh, of course. You do mention him from time to time, but always in a pooh-poohing fashion. This is what always fascinated me about your particular neurosis—this whole denial thing. I mean, it wasn't just that you took a skeptical attitude toward the UFO phenomenon and its resultant twists. It was as though you had a personal vendetta against it—or against something that happened to you.''

''That's nonsense.''

''Is it? I have your file, Scarborough. We know you better than you know yourself. We have your pyschological profile, your history—a real study of Dr. Everett Scarborough, the man and the machine. Why do you think we were so successful in pulling your strings.''

''Why are you telling me all this? You know I'll just use it against you if I can. And if you're going to kill me, you're just wasting your breath, don't you think?''

''We don't want to kill you. We want to know you . . . better. And the knowledge you will gain now,'' she smiled mockingly and imitated the flushing of a toilet, ''into the sewers. And don't think we can't do it. So relax if you fear for your life. Cooperate. This will all work out . . .''

''Why do you associate me personally with this alien abduction? And if you can erase my memory as you claim, why don't you just go ahead and tell me why you've been encouraging it, if not perpetrating it.''

''I'm asking the questions now, Scarborough,'' she said, all cold business again. ''Maybe a little drugged hypnosis might drag it out of you.''

''If you'll give me some evidence, maybe I can respond!''

''Have you had . . . odd dreams?''

''Of course. Hasn't everyone?''

''Not a good question. Let us phrase it this way. You've studied the abduction phenomenon. You've written about it. You must be aware that memories often pop up long after the supposed encounter with strange beings. Dreams, flashbacks, visions—whatever. In fact, I believe one of the principal writers on the subject . . . what is his name? Ah, yes. Maximillian Schroeder. At the end of his first book, he lists these symptoms, and he got quite a response from the populace. Have you had any of these symptoms, Scarborough?''

The word came unbidden to his mind:

Lately . . .

He dammed it up. He refused to acknowledge it.

"No, not at all."

"Hmmm. What about this period of blackouts, back in your college days. We have close to no information on them except that they were apparently not caused by drugs or alcohol."

Blackouts. The word triggered an instant response in Everett Scarborough. Tension. Denial. That was a period in his life he had thought he'd successfully forgotten. And yet, here was this CIA doctor with knowledge of the college blackouts, the lost days at MIT, regurgitating that time to him as dryly as she might discuss a bed-wetting period in his childhood.

"Your research *is* thorough," he said, "Yes, there was that odd time. Back in my junior year, working on my bachelor's. I consulted a doctor—yes, of course. That's where you would have found out about it . . . Doctor's records. There's nothing more or less than what's included there. A few spasms of the brain. They've never been repeated."

Cunningham bit her lip. "One was for three whole days, as I recall."

"Yes, but I wasn't missing. I attended classes, I socialized. I just couldn't remember any of it later."

"But there *were* times when you were missing."

"I lived alone in an apartment. I could have been there."

"However, you *don't* remember—that's just the point."

"And so you're saying that I could have been in a flying saucer?"

"No. I'm saying that this could have been part of the phenomenon. Dr. Scarborough, I have made a life's study of the brain, and to say that it is a mystery wrapped in an enigma is simplifying matters. However, thanks to the facilities presented me by my position, I have learned things far beyond the normally accepted common knowledge. You are quite correct in your suppositions that everything can be explained logically—however, I believe that your so-called logic is much too finite."

"So what does this have to do with me having an abduction experience?"

"That's what I mean to find out."

"You're not going to find anything, I can promise you that," Scarborough snapped.

"Well, we've plenty of time to find out, haven't we. You

know, Doctor—as you might have guessed, I also have your medical records." A slightly puzzled expression passed over Cunningham's face as she stepped over to an open folder on the table. She studied it for a moment, then tapped it. "A normal medical doctor might not notice some of these things. The blood-type for example. Very unusual. Curious enzyme functions. A few other tests are curious . . . oh, maybe not by themselves, but all taken together, they paint a picture of great interest to a specialist in body chemistry. Some of your brain readings, for example veer dangerously close to schizophrenia on a chemical level—and yet psychologically you betray absolutely no sign of that mental illness."

"I'm glad to hear that."

"I'm sure you are. But tell me a little more about your early life, Everett Scarborough. Have you ever felt . . . different? Like you didn't *belong*?"

"I thought that was a normal phase that every adolescent went through."

"Was that when you felt it?"

"I must say, these are very strange questions. I thought you were going to going to try to prove that I've stolen U.S. defense secrets."

"Hardly. We're perfectly aware of your exemplary record. And frankly, your reaction to the discovery of the deception that has been perpetrated upon you by our group is predictable to a certain extent. You anger, your paranoia . . . It all generally falls within the boundary of your psychological profile. And you are not a coward, Dr. Scarborough." She smiled slightly. "However, I must say, you have caused great chagrin to my superiors."

"Your superiors have caused great chagrin for me."

"And so they have." She turned a page of the report, studied it for a moment, speaking even as she read. "So you say that as a teenager you felt alienated—not one of the crowd, so to speak? Have these feelings ever cropped up again?"

This line of questioning was irking him, but he had to keep the conversation going. He had to buy himself more time. "From time to time I have felt at odds with the universe. But I don't understand where this direction of discussion is taking us?"

"Leave that to me, Dr. Scarborough. I'm not asking these

questions for your benefit. Besides, I only have one more. Tell me, Scarborough. Have you ever felt as though you didn't quite belong *here*?''

''Belong where?''

''On this planet?''

''That's ridiculous. I'm a human being. *Homo sapiens*. Of course I belong here. Where else *would* I belong?''

''You tell me, Scarborough.'' She went over to a cabinet and pulled out a drawer. From the drawer she lifted out a tray containing a number of vials and phials. These contained fluids of various colors, though the majority were neutral. ''You tell me.'' She pulled out another tray, and Scarborough recognized the glint and gleam of metallic instruments, the sparkle of graduated hypodermics.

''I don't know where you're headed,'' he said desperately, trying to attenuate the conversation.

''Not consciously, perhaps. Let's just say, I'm priming the pump.'' She lifted a large hypo up and examined it against the light. ''I strongly suspect that your subconscious understands, eh? Hidden memories? The dim, forgotten past? That, Scarborough, is what we're going to probe next. A bit of dredging, some adjustments . . . You'll be a whole new person! *That* process is my speciality.''

''The Inquisition of Torquemada had its rack and its torture helmet and its burning rods. You have your chemicals.''

She put down the hypo. ''But don't you see, Scarborough. It's all chemistry. It's the secret of life, the secret of the mind, the secret of just about *everything*—on the human level, anyway. You drink a cup of coffee when you get up in the morning—you're trying to adjust your chemistry. Imperfectly, mind you. You have one of your George Dickel's—and by the way, Scarborough, you really should stop . . . we've got some liver damage recorded here—you're trying to alter your chemistry. Instinctively, we all know that it's the secret—we just don't know how to deal with it. That has been my study, Scarborough. And I have solved some of the riddles. Ruthlessly? Perhaps. Maybe I have my own reasons for that—my own demons. You know the saying, 'Physician, heal thyself'? Well, perhaps this physician *has*.''

''Why are you telling me all this?''

She shrugged. "Someone to talk to. A captive audience with the intelligence perhaps to understand. Maybe you fascinate me, Scarborough. Maybe I'm using you as some sort of father-confessor. What do you think about that?"

"I think that could be true. But you're such an intelligent woman—don't you ever have qualms about what you do? Don't you have any respect for human rights, human dignity?"

She looked at him and suddenly her eyes were cold flint. "Maybe I used to, Scarborough. Maybe, as a young girl, I actually even had high-flown beliefs. Ideals. Maybe I even marched in a few protests, joined Greenpeace, what-have-you. But you know, as you grow up you start realizing the truth. And that truth is that human beings are horribly disfigured creatures inside. There was no respect for human rights and human dignity in the people I met, Scarborough. The people who hurt me, who formed me." A curious half-crazed smile lit up her face. "Perhaps that's what I'm after. To perfect the human race. Better living through chemistry! Compassion and love, through a needle." She picked up a hypo and regarded it. Then she put it back. "You don't look very comfortable, Scarborough. Maybe we're going to need a little assistance from the men outside to make you so."

"The handcuffs aren't enough?"

"Somehow, I don't think so." She strode to the door, opened it. "Gentlemen, would you come in here? I want you to hold the patient while I give him a tiny injection. It won't take but a moment of your time."

One of the men smiled. "Hey, Doc. It's a hell of a lot more interesting than guarding the goddamned door."

The other chuckled. "Yeah. If I'd wanted this kind of detail, I'd have filed for a duty at Leavenworth." All very light and jokey, as though this were all part of some Company sitcom. Apparently, now that Richards was away and the prey was securely fastened to a solid object, they felt they could relax.

However, Dr. Julia Cunningham was not in the mood for either a comedy or jokiness. All it took was an icy glare, and the suits were abruptly quiet, all business again. They took their places beside Scarborough, each grabbing a shoulder.

"Williams, would you be so kind as to roll up Scarborough's sleeve?"

"Look," said Scarborough, desperately. "Is this really necessary? You're an intelligent person. We can talk about this. We'll work on this together." Tersely. Not showing fear. "You've convinced me of the efficacy of your work, Dr. Cunningham. But couldn't we try it this time without drugs?"

"Nonsense."

"Doctor, the sleeve's caught up in the handcuff," said one of the men.

"No problem." Metal clattering as the woman reached onto the tray. Stainless steel flashed as she brought up a scalpel. She stepped forward, motioning the man to step aside. "I suggest you stay very still, Scarborough." She grabbed his sleeve and slashed his sleeve up to the shoulder. Scarborough could not help but wince as the sharp, cold metal kissed his skin.

"Very well," she said, stepping back. "You may resume your position."

Just then, however, another man entered the room.

Scarborough saw him first peripherally, and could not help but react: it was Myers. Edward Myers, walking into the room, quietly and determinedly. Scarborough struggled fruitlessly at his bonds, knocking against the agents. They were forced to grab him and hold him.

"Myers!" cried Scarborough. "Myers, you son of a bitch! How could you *do* this to me?" he screamed.

"Myers," said Julia Cunningham, frowning. "What are you doing here? I thought you had a plane to catch."

"I missed it, I'm afraid," said Myers, moving over with an agile step. "So I thought I'd take care of an outstanding matter."

He pulled the gun out with the practiced ease of a professional, stepped behind Dr. Julia Cunningham, grabbed her arm, and placed a silenced muzzle against her forehead.

CHAPTER THIRTY-SIX

''All right, gentlemen. I want you to take out your weapons and throw them to the floor.

''Jesus Christ, Myers!'' said one. ''What the hell do you think you're doing?''

''Making amends, Gordon. Now do it, or I'll fucking kill her!'' Something akin to madness danced in his eyes. ''I'm getting you out, Scarborough. I never should have taken you here. I'm sorry.''

The suits looked to Cunningham. She was stiff as a board, nothing but a deadness in her eyes. No fear. Nonetheless, she nodded to the men for them to obey.

Automatic pistols hit the floor, sliding over against the cabinets.

''All right, Doctor. Now, where are the keys to those cuffs?''

''The pocket of my lab coat,'' she said, monotoned.

''Great. You wanna get them out for me and free this much-abused man in that chair there?''

She nodded, and reached into her pocket. Even as he saw the glimmer of stainless steel, Scarborough remembered: Cunningham hadn't put the scalpel away.

''Myers!'' he cried. ''Watch out. She's got—''

Metal flashed as Cunningham pulled the scalpel she'd hidden in her pocket up and slashed viciously across Myers's left arm and chest. With a gasp of alarm and pain, Myers released her and she leaped away from him.

''Kill him!'' she shrieked. ''Kill him *immediately*!''

The CIA men dived for their fallen guns.

However, the wound, though painful and startling, and now

quite bloody, was far from fatal, and as Myers still had use of his gun hand, he lifted the silenced weapon toward his colleagues. The gun coughed. Swung. Coughed again. A grunt, a groan, and the CIA men toppled, bloody holes torn in their neat grey suits. Cunningham squealed with frustration and made for the guns herself.

Myers turned, aimed, and squeezed off another round.

The bullet struck Julia Cunningham in the shoulder. She toppled, striking her head against the metal cabinet, and oozed down onto the floor, unconscious.

"Glad I thought to put a muffler on this thing," said Myers. "MPs and my compatriots would be swarming all over the place by now if I hadn't. My God, that bitch has claws, doesn't she?" He looked down his arm, wincing. Scarborough could see the wound was bleeding, but apparently the scalpel hadn't cut anything major. Myers stuffed a towel under his shirt.

"Ed," said Scarborough, still stunned. "Why?"

"I fucked up, pal. I let them intimidate me. They threatened my family and I guess the old protector bullshit kicked in, cause it was fucking Richards and I knew that Richards meant what he said." He sighed and walked over to the fallen doctor. He patted her pockets. They jingled. He pulled out the keys. "But you know, in the end, a man's alone. There's nobody else in the coffin when they bury you, and you've got a long time to stew in what you've done. And I couldn't live—or die—with what I'd done to you, Scarborough. I'm just going to have to take the rest as it comes."

He walked over and unlocked the cuffs on Scarborough's legs.

"I've got a friend over at the Officers' Club at the main part of Kirtland," Scarborough said. "She's waiting for me. We'll get protection. We'll expose this whole mess. And Myers—my friend, *I* shot these guys, okay?"

"Thanks, but that probably won't wash." He unlocked the wrist cuffs. "That sounds good, though. This whole thing's the most ungodly pile of stuff I ever heard of. When it gets exposed, I think it's going to make Watergate look like a bubble bath. We'll see where things fall then."

"You'll testify?"

"You bet, Ev. I'm not done redeeming myself. I've got some serious stains on my soul. I've—"

The sound was like the crack of a whip. Myers arched his back, his face going totally white. He staggered and fell onto Scarborough.

Over his shoulder, Scarborough could see Julia Cunningham getting up. She held one of the men's guns in her hand. Blood runnelled down her white lab coat from the wound in her shoulder. Her blonde hair had come undone and it hung around her head now like a mangled halo.

"Don't do a thing, Scarborough. Just stay there. I'm calling for help."

"The hell you are," said Myers suddenly, rolling around and bringing up his gun.

He shot her in the chest and the impact of the bullet turned her around and flung her onto the trays of drugs and instruments she had laid out upon the top of the cabinet.

Julia Cunningham screamed.

She groped for purchase, but found none. She slid down the face of the drawers, and the trays tumbled with her, spilling broken glass, clamps, forceps, scalpels, hypodermics, and the sparkle of released fluids directly on her face and her wounds.

Julia Cunningham screamed again and began to spasm.

Myers had taken himself off of Scarborough, but now he collapsed onto the floor.

Julia Cunningham still spasmed.

Scarborough could see now that not only had she pierced herself with several instruments and broken vials and phials, but a large quantity of the psychoactive drugs had clearly entered her system and were kicking in with a vengeance.

Not only was Dr. Julia Cunningham dying; she was dying on the worst drug trip imaginable.

She gasped and spasmed. Her body warped into a distorted pretzel's parody of a human being and her face contorted into an animate rictus. Her eyes bulged almost to bursting from her sockets and a purpled tongue wiggled from her mouth like some fat worm trapped between her teeth trying to free itself.

Blood gurgled from between her lips.

Scarborough turned away from these dreadful floppings and twistings to attend to Myers; Cunningham could do no harm now. Myers lay on the floor, breathing slowly and shallowly. Scarborough knelt down by him, taking the gun from his hand.

The man's eyes fluttered open, gazing dimly at Scarborough as though from a far distance. "This is best . . . Family . . . safe now . . ." he said in a low voice. "Scarborough . . ."

"Yes, Myers. What is it."

"Camden. Down the hall. Turn left. Four doors. I left it open. Your friend, Jake Camden. Not in good . . . shape . . . but better than I am." He coughed up a gout of blood. His eyelids trembled and it looked as though the little light left in the eyes was guttering out.

"Myers . . . no!" Don't . . . wait . . . You have to tell me . . ."

"Tell you?" A slow muttering of lips.

"Diane! My daughter. Where's Diane!"

The head shook slightly. "They never took Diane, Ev. That wasn't the CIA, it wasn't . . . the Editors and Publishers . . ."

"Jesus Christ, then what happened to her . . . ?"

The last words gurgled up from Myers's lips like the amen to a benediction. "The Others have her, Scarborough. The Others . . ."

"Others? What the hell are you talking about, man?"

But the lips were still now, the light all gone from the eyes.

Edward Myers was dead.

"Goddammit!" cried Scarborough. He looked up. Julia Cunningham was still feebly moving in the spray of glass and drugs and blood, her hands clawing at her face.

Scarborough got up and went to her.

"The Others, Dr. Cunningham," he said between her moans, kneeling by her. "Who are the Others?"

She opened her eyes and there was nothing but fear and horror in them.

"Get away! Get away from me!" she gasped.

"The Others, woman. *Who are they? Where are they? Where is my daughter! Who are the Others?*"

"You are, Scarborough," she said in less than a voice, more of a croak. "You are."

And then with one final racking paroxysm of pain, the woman died.

The woman had been crazed with the drugs. She wasn't making any sense. Nonetheless, there was something he could do, and he did it. He went to the top of the desk where he had seen her poring through that sheaf of papers that was the report on him. He folded them up and stuck them in his pocket.

Myers was right, of course.

He had to get out of here.

He found Camden exactly where Myers had told him he would be. The man lay upon a table, looking as though he had just been through the bender of his life, and was about to check out of this veil of tears for good.

"Camden! Camden, man! Are you all right?"

"No. No, I'm not."

There had been no sirens in the halls, no alarms, no running of booted feet. In fact, Scarborough had seen nobody. This part of the installation must be relatively understaffed. That made sense, considering the top secrecy of what went on here.

This was good. It meant that they had a chance to get out. If he could get Jake Camden here moving.

"How about a glass of water?"

"How about a glass of whiskey?"

"Water will have to do."

Scarborough went down the hall, found a toilet which conveniently included a drinking-water facility complete with paper cups. He poured out two. The first he fed to Camden slowly. The next he dashed into Camden's face.

"What'd you do that for?" Camden said, spluttering.

"Come on. No more time for fooling around. If we don't get out of here now, I'm afraid we're going to be stuck here for a very long time."

That got Camden moving. "You're right." Weaving a little and clearly still not entirely oriented with the real world, Jake Camden got up and started wobbling toward the door. Scarborough grabbed his arm and steadied him. "Glad to see you've got that piece, though."

"Too bad I can't shoot very well."

"You want me to take it?"

"Camden, thanks, but I'm afraid you'd accidentally shoot *me*!"

"Maybe you're right."

"Now the question is, how do we get out of here."

"You want to take a look at these facilities first? It'll explain just a *hell* of a lot to you about what they've been doing."

"You describe them to me later. Right now, I just want to get out of here."

"What about Diane?"

"Myers says they don't have her."

"What about Myers?"

"Dead."

"Shit. And that blonde doctor?"

"Very dead."

Camden grunted. "I think I feel lots better."

"I think she does, too."

They hurried down the hall, not quite knowing where they were going but wanting to get there fast.

CHAPTER THIRTY-SEVEN

The wind blowing through her hair, the map of the base flapping on the seat, the borrowed Toyota Celica smoothly eating up the miles, Marsha Manning drove toward the CIA installation at the far end of Kirtland Air Force Base.

Of course, it wasn't marked as such on the map. In fact, it wasn't marked at all—the only indication of buildings on the periphery of Kirtland were unmarked blocks. For all the map-reader might know, they could just be outhouses. There was absolutely nothing else close by, so it had clearly been the perfect place to house a secret operation.

Marsha just wanted to get there, as quickly as she could. By now, there was absolutely no question in her mind that Scarborough had run aground. There was no other explanation for his not getting to the Officers' Club either on time, or late. She'd even given him the phone number that she could be paged by: no call, either.

And she kept an eagle-eye on the road for vehicles going the other way. This was the only road to the place, and the whole half hour she was on it, not one car or truck passed.

She just prayed that he wasn't dead.

Somehow, though, she couldn't believe he was. Whatever bond that they'd developed over their time together didn't feel broken and Marsha knew, she just *sensed*, that he was still very much alive and with her.

He was just in deep trouble.

When she arrived and saw the fence and the gate, she knew there were going to be problems. But she'd suspected that there would be, so she had a plan of operations.

"Good afternoon, Lieutenant," said the airman first class at the gate after the appropriate salute. "May I see your identification and your pass?"

"Certainly." She gave him her ID, which he scrutinized much more carefully than they usually did at these points.

"And your pass."

She handed him her specialist's card, the one identifying her as a top computer systems analyst for the Air Force.

"That's very well and good, but that's not what I need. Who are you visiting here?"

"The computer people. They've got some sort of emergency, and they just sent me out to take care of it."

"I've no notification here to expect you, Lieutenant Manning. In these situations, that's the usual procedure."

She sighed. "What, am I going to have to go all the way back to the main base to get something in *writing*."

"Do you have a name I can call?"

"Yes." Desperately, she wracked her brain for one of the names that Scarborough had used. "Doctor . . ." she began, not quite sure what might follow. "Doctor Julia Cunningham. Please call her and she'll come out *personally* and make sure I get in. Will that be sufficient?"

The guard did not smile or even nod his head. He just went back into his booth, and began to page through a book. He lifted a phone from its cradle and commenced dialing.

So what does the heroine do now? she wondered. Put the pedal through the medal and crash through the gate post? Not especially smart, really, as they weren't exactly making them out of easily breakable wood anymore, but rather of a quite sturdy alloy. No, the thing to do was just to wait, and bluff if she could.

Yes, wait it out. Play it as it came.

About a hundred yards past the gate was a squat and spartan-looking building. From a side door two figures were emerging, and Manning immediately noted that they definitely did not look either Air Force or CIA.

Indeed, they looked rather ragged and furtive. They began walked the opposite way as though they were looking for something.

"I'm sorry, Lieutenant. But no one seems to be answer Doctor Cunningham's line. I'm afraid that—"

"Airman, wait, I see a man I'm supposed to be working with. Walking out of that building over there. I don't especially want to go all the way back to the main base. Is there any way that we can contact him?"

The corporal looked startled. "That's highly irregular."

"Airman, I happen to know that bullhorns are standard issue for guard gates. You could use that. Quickly."

"I've never—"

"Airman, I appreciate the confidential nature of this installation. However, we are presently actually on military property, and I am your superior officer. Consider this an order!"

The airman reached down and pulled out the bullhorn. "What's his name?"

"Captain . . . Captain Everett. Mention my name."

The airman put the horn to his lips, flicked the switch and projected his voice out into the compound.

"Calling Captain Everett. Attention, Captain Everett. Lieutenant Marsha Manning requests your assistance at the gate."

Marsha could see the two figures turn around. For a moment, it looked as though they were about to bolt . . .

No, Everett. No, Everett, don't! It's me! Me!

. . . but they did not.

Instead, after a moment of conference, they started walking toward the gate.

"Thank you, Airman. I'll make sure to commend you to your superior."

"Wait a minute. He's not in uniform! That's no captain. I recognize him. That's . . ."

Marsha Manning took the gun from her purse and aimed it at the airman. She hadn't wanted to do this, but this was a definite situation involving no choice.

"If you'll turn around, you'll see that it would be best for you to shut up and keep your arms where they are."

The guy turned and his eyes started from their young head. Suddenly, at the sight of the bore of a gun directed at him, all his training drained away. He tried to say something, but apparently could not.

"Good. I want you to stay exactly like that, or I'm going to have to kill you." A strong threat should be enough to keep him quiet for the minute or so it took for Scarborough to get there.

Whether they sent her to Leavenworth or not, her Air Force career was over now, anyway.

So what the hell?

". . . Attention, Captain Everett! Lieutenant Marsha Manning requests your assistance at the gate."

When the first electronically boosted words broke like distant thunder, the first impulse Scarborough had was to run. Camden, recovering rapidly, was right on his heels. But the man grabbed Scarborough by his arm and hauled him back to a halt.

"Hey, man! I think they're calling you!"

That was the precise moment that the "Marsha Manning" part of the message sunk in.

He turned around.

There, some distance away, a blue Toyota Celica stood in the shade of the guard-post awning. Scarborough's eyes were good, but not that good—he could not see who sat in the car. But he *could* see that it was a female by the fall of long hair.

Marsha! She'd come for him after all! And there she was, right on the edge of escape.

All they had to do was to get there.

"Come on!"

Camden did not have to be prompted further. In an incredible burst of speed for a man in his state of health and mind, he virtually loped ahead of Scarborough's more stately run.

Ah, for the energy of youth, thought Scarborough as he gamely increased his speed.

The race proved to be a tie, with Camden's energy all but

shot by the time they reached the car. An infusion of joy almost overwhelmed Scarborough when he saw that the woman was indeed Marsha Manning—and that she was holding a gun on the guard, preventing him from calling for help.

Camden staggered to the other side of the sedan, opened a back door and flopped into the back seat.

"Marsha! Am I glad to see you!"

"Don't get too excited. Who's the guy in the back seat?"

"Jake Camden."

"What happened to Walter?"

"Dead. A trap, not far from here. Tell you later. Right now, we gotta go."

"You're Scarborough," muttered the guard. "You're the guy out here . . ."

"That's right, my friend. Did you enjoy the show?" Anger bubbled up in him like a tide. He took the gun from Marsha Manning and struck the airman hard on the back of the neck. The man went down hard, but Scarborough caught him and propped him back up in the booth.

"What did you do that for? You've just assured my court-martial!"

"He'll be all right. I just don't want him making any calls any time soon." Scarborough swung around, checked the periphery. He saw a couple men walking between buildings and a loading vehicle moving in the far distance and that was all. That was one of the fortunate things about this place—little activity, few people. They had a chance to put some miles between them, but in order to do so, this guy in uniform here had to be put out of commission, unobtrusively. Back in the booth there was a stool. Scarborough propped the man on the stool. He unplugged the telephone and used the cords to tie the man's hands and feet.

"You got a handkerchief?" he asked Manning.

Marsha fished one from her purse. Scarborough used it to gag the man, making sure his nostrils were free so that he could breath.

"Let's just hope he stays out for a while," said Scarborough. He ran around to the other side of the car, opened the door and got in.

"Okay," he said. "Back this thing up and let's get *out* of here."

"I'll second that motion," groaned Camden from the back seat.

CHAPTER THIRTY-EIGHT

When Marsha hit the main roadway, she pushed the accelerator all the way down to the floor and pushed the Celica as hard as it would go. They drove through the arid and deserted part of the base, a rooster-tail of dust trailing in their wake, as though the devil himself were chasing them.

Which wasn't far wrong.

The problem with escaping from an Air Force base in a car in a wide-open place like New Mexico is that, once the Air Force finds out (or rather the CIA with Air Force facilities), it's simply a matter of dispatching a few of the many helicopters doubtlessly on hand, bristling with hypertrained men and weapons, and either bringing you back or blowing you off the face of the earth.

Scarborough was more than aware of their dilemma.

He knew that to have any hope of escape, they had to get to a populated area as soon as possible. They had to ditch the car as soon as possible as well—but preferably not before they made it to the populated area. The latter would have to be Albuquerque. No other place close enough would fit the bill the way Albuquerque would.

According to the map of Kirtland Air Force Base that Marsha Manning had been smart enough to bring along, there was a gate at the southerly end of the base much closer to the CIA installation—and probably safer in the bargain.

It was toward this gate that Marsha now sped.

"I thought that something had gone wrong. I just had to come out, even though you told me not to."

"Thanks," said Scarborough. He was glad for the period of silence that had passed between them. It gave him time to catch his breath.

"You're not going to scold me."

He took a deep breath. "I do believe that whatever stubbornness and pride left in this aging body has been thoroughly trampled out."

"I'm so sorry to about Walter. And Myers . . . He betrayed you? And where did Camden come from?"

Scarborough told her the whole story as best he could in the ten miles or so they had before they reached the south gate. When he finished, it was in view, about a mile away.

"My God. That's incredible," she said. "That doctor . . . She sounds like she was just *awful.*"

"Bad chemistry, maybe," said Scarborough. "We'll have time to go through it all later." He hadn't told her the business about the aliens and his supposed connection to them. That was a little too much to deal with right now—he couldn't even *think* about it.

"I trust you have the necessary identification?"

"You bet."

"I'm afraid we don't."

"You don't really need ID to get off a base. In fact, I wouldn't be surprised if he just waves us on through."

"I'm sure we don't look so hot."

"You look serviceable enough, though. Your friend in the back seat, though . . . wish I had a blanket to throw over him."

Scarborough angled around and spoke to Camden.

"Jake, I want you to sit up straight and stay in the part of the seat farthest away from the guard. I want you to look straight ahead. If we are stopped, I don't want you to move, I don't want you to talk, I don't want you to do anything but stay absolutely still and act as though you are in perfectly normal shape. Do you have that?"

"Uhnnnnh," said Jake, mouth mashed against the black vinyl of the back seat.

"Jake . . . Get up!"

Reluctantly, Jake got up. Scarborough smoothed down some of his fright-wigged hair and straightened his shirt the best he could. Thank God the guard wouldn't be able to smell the guy—poor Jake smelled absolutely awful.

"Guard post, dead-ahead," said Marsha. She hauled in a deep breath and braced herself for anything as she drove toward it.

The guard—another man in the usual uniform—flagged them down, signalling them to halt.

"Well," muttered Marsha *sotto voce*. "At least he's not waving his gun."

"Doesn't look real upset or worried. Wonder why he'd making us stop though."

She slowed down in front of the gate.

The man leaned toward them with a smile. "Sorry to stop you, Lieutenant. Can I see your ID please?"

"Is there a problem."

"Yes. I don't see any ID sticker on the car."

"Oh, I see. Well, actually, it's a friend's car—not mine."

"Don't worry, I don't need registration or anything. I just wanted to have a look-see." He looked down at the typed information and the photo, then at Marsha. "Yes, everything seems to be in order. You're from Wright-Patterson, huh?"

"That's right."

"I used to be stationed there." He looked around at the vast expanse around him. "Whew, and I thought it was boring *there*!"

Manning laughed breezily. "Yes. I know what you mean. Oh well . . . is that all, Airman?"

Right on the word *'all'* the phone in his booth began to ring.

"Ooops. That's unusual. Would you just hold on for a moment, Lieutenant? I want to ask you one more thing."

The phone rang again.

Were they calling to seal off the gates? What else could it be? Scarborough could feel his muscles involuntarily tense up, tight as coils.

"Really, Airman, we are in a bit of a hurry."

"Oh."

The phone rang again.

"What was the question?"

"I just wondered how long you were going to be in New Mexico?"

"About a week." She smiled at him. "These are my two

brothers here and I'm showing them around. Why, would you like to have a drink or dinner or something?"

"Actually. . ."

"You can reach me through Colonel North's office. Call me tomorrow, okay?"

"Wow. Sure!"

Ring.

"Can we go now?"

"Oh, yes of course." A bit dazed and happy that he had a date, the airman marched to the booth and hit the switch to let the gate up. Even as it was lifting, Scarborough could see him picking up the receiver.

Marsha Manning, however, wasted no time. As soon as there was space to slide under the gate arm, she did so and immediately picked up speed.

"Camden, look behind and see what's happening."

"You said not to move."

"Orders changed."

"I'm not sure I *can* move now. I'll all stiff, man! And I think I'm gonna pass out anyway real soon."

"Well, do it now then, so I can have a line of sight."

Camden collapsed onto the seat.

Scarborough looked back.

The guy in the booth was still on the phone.

A false alarm? Surely, if the message had been: Check all vehicles for Everett Scarborough, the man would have come out, blasting away at them with a rifle. Or did that only happen in movies?

In any event, the way Marsha was moving, the booth was soon only a speck in the distance, and they were headed for open highway.

"I don't know if he knows now if it was us or not."

"Oh, if they didn't know—they will soon," said Manning, hands gripping the steering wheel, chin jutted out determinedly. "We've got to make tracks."

"What we could really use, though," murmured Scarborough, "is a different car. They're going to put the make on this one real quick."

"Oh, right, I'll just pull in at Joe's Used Cars up the road and make a quick switch."

"Just get us to Albuquerque, Marsha. That's all I ask."

"And you'll take it from there?"

Scarborough jerked a thumb to the back seat. "Well, we know that he won't, don't we?"

They hurried on down the black macadam ribbon that was the New Mexican road, bisecting the arid plane.

They saw the Winnebago Recreation Vehicle after they turned past an outcropping of rock. It was half on the shoulder of the road, and half jutting into their lane of the two-lane blacktop. A flare was stuck in the road beside it, casting off a glare and dark smoke into the clear afternoon.

"Looks like they had a flat tire or something," said Marsha, touching her brake slightly, slowly.

"Yes."

"Should we slow down and help, or go around them?"

"You're thinking the same thing I am, aren't you? You're thinking, what if this car is developing serious motor problems. What if we stop and help them with their flat, in return for a lift into Albuquerque for proper help?"

"Well, the notion *had* crossed my mind. That way, we could just abandon the Celica. The Air Force would find it soon enough and my friend will get it back. Although by that time I don't think he'll be a friend any more."

Scarborough grunted. "Looks like we stop then."

Manning pressed harder on the brake, squeezing the Toyota down to a rolling stop as she turned onto the shoulder. She turned off the ignition and they both got out of the car, squinting into the sunlight.

There were two men working on the back of the Winnebago. From the looks of it, one of the tires had blown out. They were working on jacking the heavy vehicle up using one of its strong bottom supports—but they seemed to be having some difficulty with the hydraulic jack.

"Hello there!" called Scarborough. "Couldn't help noticing your problems. We've got some of our own. But can we help you first?"

There were two men working on the Winnebago. They turned around to face the new arrivals.

One was a younger man, perhaps in his late twenties. He

wore a green-striped Lacoste shirt, blue jeans, and tennis shoes. The other, an older man, was dressed in an open-necked yellow-and-red Hawaiian shirt and white tennis shorts.

"Why yes, Mr. Scarborough," said the older man. "Yes, you most certainly can!"

CHAPTER THIRTY-NINE

His head hurt.

His head hurt bad, the sharp pain of it like a corkscrew jammed up the back of his neck.

Scarborough woke up. (*Woke up? Had he been asleep?*) There seemed to have been no jump cut of time, and yet, when he thought about it, he realized that there indeed was, yet he was so *rattled* he didn't know what side of the jump he was on, or where he'd jumped *from*.

Scarborough looked up. He realized that he was sitting in the passenger seat of some kind of vehicle, and he was high off the ground. The sunlight was bright beyond the large windshield.

He felt fear deep in his gut. A sense of doom pervaded.

Got to get out, he thought. Got to *get out of here*.

He reached for the handle of the door. It was locked—it did not move. Desperately, he banged against the door, pushing hard on the handle.

No results.

Feeling a desperate kind of claustrophobia, Scarborough lunged across to the driver's seat.

An arm reached out from behind him and powerfully forced him back into the passenger seat.

"Wha—"

"Do not move, Everett Scarborough," said a deep, menacing voice. "Do not look behind you."

Scarborough was far too stunned to do anything.

"Who are you?" he managed. Which was really *stupid*, he thought, because he knew perfectly well, deep in his gut, who it was.

It was *them!*

"That is too early to reveal."

"You've got Diane . . . You've got my daughter."

"That is correct, Everett Scarborough. Remember that. We have your daughter. Nothing will happen to her, no harm will be visited upon her—on the condition that you cooperate."

"What do you want me to do?"

Scarborough smelled Brut cologne.

There was silence. A deep, reverberating silence, filled with a profound strangeness. Scarborough sensed that the person (persons?) behind him had an uncommon *depth*, an intelligence . . . Yet how could he know that, he found himself asking himself . . .

However, there was no time to dwell on that thought.

The voice was deep and thorough in its diction.

"In the glove compartment of this vehicle, you will find a cassette player. Your instructions are on that tape."

"I see." Scarborough was tempted to jump from his seat and attack the man behind him. Pound the truth out of him. *Force him* to take him to his daughter. However, the sense of menace was so great that Scarborough had to use all of his willpower to merely control his fear. "My friends . . ."

"Your companions are safe."

"You were waiting for us . . . you were the people in the Winnebago."

"Yes. Scarborough—Do as the tape says. I cannot urge you strongly enough to DO AS THE TAPE SAYS."

"Who *are* you? What do you want? What's going *on . . . ?*"

But there were no answers.

Scarborough felt a sudden jab of pain in his left arm, a strong scent of masculine cologne . . .

And the darkness rushed in like cold, black water.

Everett Scarborough awoke in stages.

The first stage was a very long, drawn-out period punctuated by stirrings of awareness and sensations of *movement* amidst a long, long period of total blackness.

And then, abruptly, there came the dream stage.

He dreamed of falling. He dreamed of flying. He dreamed all the dreams he had dreamt before in his life, the personal ones and the classical ones—as though they were all running before his mind's eye in review, like a man's life is supposed to, before he dies.

He dreamed his nightmares.

Nonetheless, when Scarborough hit the next stage of his awakening, he did not feel frightened or discomfited. Rather, he felt relaxed and at peace. This next phase was a gentle drifting in softness, a drowsing, a languorous time floating in a sea of unconcern.

Vaguely, he was aware that somewhere along the line, the sense of motion had ceased.

He drifted off to sleep once more.

When Scarborough awoke again, he was instantly alert. He lifted his head and he looked around him.

He lay upon a bunk bed, set in the wall of a small room containing a plentitude of cabinets, a refrigerator, a table, a bar, stools and a couch, a television set, and three other bunks. The place smelled of sleep, old coffee, and recycled air. In the other bunks, lay two people, both asleep.

Jake Camden and Marsha Manning.

For a moment, Scarborough experienced a sense of total dislocation, a "where am I?" feeling magnified to the point of dizziness. But a word occurred to him, and with the word the memories rushed back in to fill the void.

And the word was *Winnebago*.

They were in the Winnebago.

He remembered now, stopping alongside the road. Getting out of the car. Asking those men if they needed help.

And then one of them had taken something out of his pocket and aimed it at them, and the memory was abruptly curtailed.

And then that time, seemingly a dream, when he woke up behind the wheel—the wheel of the Winnebago.

Was it a dream? he wondered. Most empathically, his mind answered him back.

No. It had been real. All *too* real.

His first concern was for Marsha.

She lay in the bunk set immediately above the one he'd been sleeping in. He gently shook her arm. "Marsha," he said. "Marsha, are you all right?"

She blinked her eyes and focused on them.

"Hi!" she said and reached up for him.

He couldn't help himself. He leaned over and gave her a warm embrace and a soft, meaningful kiss. It was instant relief to the anxiety that had swept over him. But nice as it was, reality took hold.

Where the hell were they?

What were they doing in a Winnebago?

Which was exactly what Marsha asked when she got a chance to take a look at her surroundings.

"I'm not sure. I'm still pretty confused myself."

"Those men—I remember now."

Scarborough turned toward the front of the vehicle. A curtain was drawn between the front and the living quarters.

"Well, let's go have a look, shall we? Maybe we'll even get a chance to talk to our captors."

He had a suspicion who they were—the men who had saved him in the New York subway, who he'd confronted in nighttime Baltimore—but he said nothing about this to Marsha. That could wait.

He helped her out of the bunk and together they went to the curtain. Scarborough flung it back, ready to face just about anything.

The only things in the front were two empty seats.

"Looks like we're alone," said Marsha.

"I'm going to check outside."

"If we're not locked in."

"Let's have a look."

He hopped over the seat, surprised at how agile and refreshed he felt, how energized. He hadn't felt so good physically since before all this began!

The handle to the door gave way easily, and the door swung out, letting in a blast of hot, dry air. Scarborough stepped out onto gravel. He walked a ways.

The Winnebago was parked on the shoulder of a highway.

In one direction were mountains and desert and clear blue

sky. In the other were eating establishments, gas stations, a used car lot, and, in the distance, the tops of larger buildings, wavered by the heat: the beginnings of a city.

"Looks like we're still in some part of the Southwest, *that*'s for sure," said Marsha, rubbing the last of her sleep from her eyes.

"Where? Arizona?"

"Maybe."

"They drove us here and abandoned us? But why?"

From the back of the Winnebago came the abrupt sounds of a pounding and clattering.

"Your friend Camden is everything you said he was," said Marsha dryly.

"He's been through a hell of a lot."

"I never thought I'd hear you defending him."

"We'd better go in to make sure he's okay."

They crawled back over the seats, into the main part of the Winnebago. Standing in the kitchen area amidst a scatter of pots, pans, and utensils was Jake Camden. He stared up blearily at them and said, "Hey, guys. Think you can help me out here? I'm just trying to get myself a cup of coffee."

Scarborough laughed, relieved. "Yes. I think that can be arranged, Jake."

"I've got the milk. This icebox is really jammed with food. But I can't find the cabinet with the coffee. And—hey. Where the crap am I, anyway?"

"You're in a Winnebago."

"A Winnebago? You mean, like an RV?"

"That's right."

A totally lost and disoriented, "what am I doing here?" look passed over Jake's face, but then he just shrugged and started looking for the coffee again.

"Make some for us, too, Jake," said Scarborough.

"What are you doing, Ev?"

What Scarborough was doing was going to the front of the vehicle.

The dream. His dream of sitting in the passenger seat of the Winnebago. It had been real.

He remembered now . . . The tape . . . The cassette player in the glove compartment . . .

He plopped down in the passenger's seat and found the glove compartment. It was unlocked.

"Looking for some kind of registration. Who owns this thing." He didn't want to alarm them, tell them of the horrible *menace* he'd felt there, the *dread*, and the *claustrophobia*.

What he found instead was a Sony Walkman.

Marsha sat in the driver's seat and stared at what Scarborough held in his hands.

"My goodness," she said. "Alice in Wonderland time."

"Well, I guess there's nothing much else to do except follow directions."

Scarborough hit the "Play" button.

For a moment, a faint tape hiss. Then:

> This message is for Everett Scarborough.
>
> Please be advised that you are sitting in a vehicle with a full tank of gas, in highly operational mode. There is a full complement of supplies in the rear, along with a change of clothing for you and your companions. Also, in the glove compartment you will not find any registration. It is doubtful you will need any. What you *will* find, however, is an envelope containing five thousand dollars in hundred-dollar bill denominations.
>
> You are presently outside the city of Prescott, Arizona.
>
> We have your daughter Diane, Scarborough. She is safe and well, and she will stay that way if we receive your cooperation in the future.
>
> Go where you will for now. It is not the right time for us to contact you directly. When that time arrives, we shall do so.
>
> Until that time, farewell, Mr. Scarborough. All questions will be answered—eventually.

Tape hiss.

"But *who* are they?" said Marsha, breaking a long moment of silence.

"Did I hear someone mention five thousand dollars?" said Camden, sticking a smiling face through the curtain.

Suddenly, Scarborough felt terribly claustrophobic. It felt as though the metal and vinyl of the Winnebago cab was constricting around him, about to crush him.

"I have to get out of here," Scarborough said in a choked voice.

He jammed the tape player back into the box, hammered open the door of the RV and jumped down onto the gravel. He strode away from the highway, toward the brush and ragged landscape. The heat enfolded him gently and the sun beat down on him like a spotlight on a vast stage.

"Everett! Everett!" called Marsha after him. "Are you all right."

"I need some time!" he called out, as though as much to his surroundings as to himself. "I need some time to myself!"

She did not run after him, and Scarborough was glad of that. Fear and emotion and turmoil burned in his chest, blooming like a terrible ravaging explosion.

He walked for perhaps a hundred yards before he came to a group of cacti. He sat down upon a flat rock. He realized that there were tears streaming down his eyes.

Diane, he thought. *Oh Diane*.

The words from Ed Myers and Julia Cunningham burned in his ears . . .

The Others . . .

The Others . . .

Did they exist?

Were they *really here*?

But if so, who were they?

And in God's name, what did they want with *him*?

He sat like that for a time, until the heat dried his tears.

And then Everett Scarborough remembered the report—the papers that he had torn from the notebook and stuffed into his back pocket. Painfully, almost reluctantly, he reached back, pulled them out, and unfolded them.

He read them.

When he finished, he carefully refolded them and placed them back where they had been.

He fell to his knees, clenching his fists and biting his lip until it bled.

When the worst of the spell passed, he stood and took in a deep breath. Another.

And then he looked out over the desert and the distant mountains and the cold, painfully blue sky, wishing that he were dead.

But no . . . he wasn't dead. He was alive. And someone still had Diane . . .

He had to find who that someone was.

He stood and determinedly walked back to his friends waiting for him at the Winnebago.